Quick, Practical Guide to *DOING* Ham Radio

Everything You Need to Know to Get Licensed and Operating!

Timothy S. Jacobson

N9CD

Cover background image ("Earth at Night"): NASA Earth
Observatory images by Joshua Stevens, using Suomi NPP VIIRS
data from Miguel Román, NASA's Goddard Space Flight Center

Cover design: Timothy S. Jacobson

Photographs in the book are by the author unless otherwise noted.
Dipole diagram and "Electromotive Express" drawing
by the author.

Published by

Driftless Press
1140 Crescent Hills Dr.
La Crescent, MN 55947

DEDICATION

This book is dedicated to all the Elmers, ham radio club officers and volunteers, Volunteer Examiner Coordinators (VECs), ARES and RACES operators, storm spotters, and other good folks, who serve, protect, mentor, innovate, build, teach, and bring joy or assistance to others around the world through the Amateur Radio Service and wonders of radio propagation.

WHAT OTHER HAMS ARE SAYING
ABOUT THIS BOOK

"Starting out in amateur radio can be an exciting but daunting endeavor for a beginner. There are many facets of ham radio, and Tim Jacobson's book provides a great deal of helpful guidance into this new world.

I've been a ham for more than 50 years, and I am impressed with Jacobson's breadth of understanding of what ham radio is about and the elements that are open to us in this hobby. I'm sure that you'll find this book a quick read, but filled with practical tips. Jacobson has done an admirable job presenting this primer into a wonderful hobby."

Michael Foerster, W0IH

"Tim Jacobson's *Quick, Practical Guide to DOING Ham Radio* will prove to be an excellent resource for both the prospective and the recently licensed radio amateur."

Mike Cizek, W0VTT

"A great book that provides all you need to know to get licensed and on the air."

Dan Abts, AB9TS

"*Quick, Practical Guide to DOING Ham Radio* is an excellent source for anyone interested in amateur radio. Tim removes the mystery of what 'ham' radio is, and how to go about getting on the air. The information given should encourage many people to explore the rewarding study of amateur radio."

Eric VanOsdol, AC9EV

CONTENTS

ACKNOWLEDGMENTS

I would like to acknowledge my *Elmers*—those experienced hams who have helped me get started on my amateur radio journey, especially Bob Seaquist (W9LSE), Mike Foerster (W0IH), Dan Abts (AB9TS), and Mike Cizek (W0VTT).

Especially, I would like to acknowledge and thank Eric VanOsdol (AC9EV), Mike Cizek (W0VTT), Mike Foerster (W0IH), and Dan Abts (AB9TS)—a great group of Amateur Extras—for their knowledgeable critiques and suggestions for the content of this book.

Also, I would like to acknowledge my wife, Lisa, who provided much encouragement as I wrote this book, and who tolerates me hunkering down in my basement shack and repeating my call sign thousands of times (which she can hear through the floor) during ham radio contest and QSO party weekends.

1

INTRODUCTION

If you're anything like me, you have way too little time to do the things you love or to explore new activities that you might have an interest in.

Amateur radio, often referred to as "ham radio," is a vast subject. It encompasses equipment, operating practices (with multiple modes of operation, such as voice, Morse code ["CW" in hamspeak], video, and digital modes), antenna design/construction/placement, electrical theory, electronics, safety, etc. While ham radio's vast array of aspects, options, and opportunities provide great joys, it also can be a bit overwhelming and create a big obstacle to getting started.

I know, from personal, *recent* experience what it takes to go from essentially zero knowledge about ham radio to operating as an "Amateur Extra," the highest class of license available in the U.S., in a short amount of time.

Unless you have a good and patient friend who is an experienced ham operator when you are just starting out—whether you remain undecided about whether to pursue ham radio or are anxiously awaiting the opportunity to dive in—the questions you have are legion and the answers can be difficult to sift through. The result can be unnecessary wheel-spinning and discouragement. This leads, at a minimum, to delayed fulfillment and the waste of precious time and money. Even worse, it leads to people choosing not to pursue the amazing and wonderful art and science of ham radio.

There are tricks and shortcuts this book will quickly teach you—so that you can jump right in and enjoy the activity, regardless of which facets of ham radio you are most interested in.

As someone who has learned a wide array of skills, including playing guitar, flying airplanes, scuba diving, and producing documentary films, let me give you a key tip:

> **Don't make the mistake of getting bogged down in trying to learn every aspect of ham radio before you allow yourself to enjoy actually doing it!**

There are some in the ham radio field who push prospective hams to learn large amounts of detailed, esoteric, and technical details before they get to experience any of the joy of actually DOING ham radio. In my view, this is, quite simply, wrongheaded.

Don't misunderstand: I am not poo-pooing the idea of learning as much as you can (or can tolerate) about things like electronic components and theory or regulatory issues. What I am saying is that, contrary to the message seemingly conveyed by the FCC, in order to get on a radio and communicate with other people, you do not need to know that a silicon NPN transistor is biased "on" when the base-to-emitter voltage is approximately 0.6 to 0.7 volts, or whether voltage leads or lags current in a capacitive circuit. (You *will* need to learn such technical details to get an Amateur Extra license, but many hams operate their whole adult lives without pursuing that particular license, and they still get to have a lot of fun!)

Ham radio is a bit like flying airplanes. As a pilot, I understand and appreciate the importance of prospective pilots learning everything they can about the physics of flight, mechanical aspects of airplanes, navigation techniques, weather, and regulations to ensure enjoyable, safe flying experiences. That being said, it is important to note that my primary flight instructor let me take the controls of a Cessna airplane on my very first flight for most of the duration of the flight. He let me experience the joy of flying before getting worried about whether I had memorized regulations about airspace designations, instrument meteorological conditions, navigation methods, etc. He recognized that it was possible and desirable for a pilot candidate to experience the joy of flying and to learn by doing and not just by reading. Sixteen years after getting my pilot's license, I still am not an Airframe and Powerplant (A&P) mechanic authorized to *fix* airplanes, nor do I ever aspire to be an A&P. In the aviation world, pilots outnumber mechanics about 2.5 to 1.

In contrast, in the ham radio world, the licensing system essentially expects 100% of amateur operators using the high frequency bands (General and Amateur Extra Classes) to be "mechanics" in the sense of having to know a lot of technical details about electronics and electrical theory. Some people in the ham radio field want prospective hams to understand the theory of running before they've even experienced the joy of walking.

If you're not an electrical engineer, don't despair. The Technician Class license with its 35 questions is *easy* for most folks. (You only need to answer 26 questions correctly—75%.) There is no CW (Morse code) requirement (for that or any other ham test these days) and all the test questions and answers are available on the Internet. It generally only costs $15 to take the test, although at the end of 2020 the FCC approved a $35 fee for each application for an amateur radio license, the timeline for which has not been established as of the time of this writing.

For a little more studying, the General Class license grants some operating privileges on *all* Amateur Radio bands and *all* operating modes, opening the door to worldwide communications! I strongly recommend that all prospective hams prepare for both the Technician and General Class licenses. Just like with the Technician Class license, the General Class test is a mere 35 questions, there is no CW requirement, and all the test questions and answers are available online. You can take both tests together for the same fee.

If you pursue a license for amateur radio, you will be in good company. Some famous people who have been hams include Tim Allen, actor; Joe Walsh, singer, composer, record producer; Walter Cronkite, newscaster; Marlon Brando, actor; Chet Atkins, singer and composer; Garry Shandling, actor, comedian; David Packard, co-founder of Hewlett Packard; John Sculley, former CEO of Apple Computer; Barry Goldwater, U.S. Senator; Jim Croce, singer/songwriter; Dick Rutan, pilot/adventurer; Priscilla Presley, actress and wife of Elvis; Yuri Gagarin, first cosmonaut; King Hussein of Jordan; and Rajiv Gandhi, Prime Minister of India.

But you don't have to be famous, popular, or rich to be a licensed ham. It's an activity accessible to almost anyone. Sometimes young people will learn CW (Morse code) so that they can communicate with adults around the world without anyone realizing (from their voice) that they're "merely" a kid. Others get licensed in their later years so that they can talk to distant friends on a daily basis. There are many folks with physical disabilities who do ham radio, too.

Do not hesitate to pursue an amateur radio license. There is no downside. You will learn interesting things, and you will learn skills that may be helpful in times of emergency. The cost of licensure is low. Free and low-cost resources are abundant. By using a remote high-frequency rig (a radio operated through an Internet interface) affiliated with a club or a friend, you might even be able to get on the air without investing in any hardware—giving you a chance to talk to people around the world and see if you enjoy the activity or not. And for as little as $25, you can acquire your first handheld VHF/UHF transceiver. There has never been a better time in history to become an amateur radio operator, which is why there are record numbers of people getting their license.

Do it, and do it now! You will open up a whole new world to yourself.

2

EARLY EXPERIENCES
THAT LED ME TO HAM RADIO

Even though I do not work in a science field, I have had a strong interest in science for as long as I can remember. I think that springs from insatiable curiosity and a profound sense of wonder—wanting to know how the world works. Certain aspects of radio have surprised and intrigued me over the years.

I remember my first experience with the amazing world of radio propagation. It was in the mid-1970s, and I was probably between seven and ten years old, living in a tiny town in northern Wisconsin. My family's Ford Galaxy 500 was equipped only with an AM radio (and, at some point, a CB radio). I can remember sitting in the car late on a summer night, parked across the street from our house. I was listening to the radio with a friend and looking out at the stars. We were picking up stations from far away—stations I never heard during daytime. In particular, I remember a station from Chicago—it must have been WLS—and the signal was coming through crystal clear. My friend, who was several years older than me, explained that the signal was skipping off the atmosphere at night, which is why we were hearing the station so well.

Probably within a year of two of that experience, I first heard hams on the air. The parents of one of my friends bought a shortwave radio. I remember tuning across the bands with my friend, also named Tim, and hearing people speaking various languages as well as mysterious Morse code tones. I don't think we spent very much time with the radio—it was just a brief novelty for us.

Another aspect of radio that caught my attention was the plan for building a crystal radio in what I think must have been the Cub Scout handbook. The idea from those plans stuck in my mind until 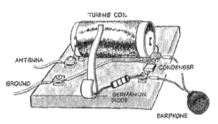 my 20s when I finally ended up building a radio receiver using scraps from a broken electric razor, a junked toaster, and a germanium diode from Radio Shack.

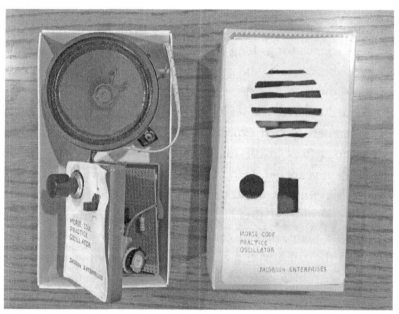

Morse Code Practice Oscillator the author constructed circa 1991 (opened up).

It was in my 20s that I first seriously considered pursuing a ham radio license. As a diversion from the grind of law school studies, I started teaching myself basic electronics. In addition to building an AM radio, I constructed a Morse code tone generator, and I started trying to teach myself CW.

I may have pursued a ham license back in the early 1990s, but I didn't know a single ham operator, I had never seen a ham radio, and there wasn't yet a World Wide Web filled with information about how to get started. In other words, there was no one to encourage me, and resources were much more limited then. Even though I did not pursue an amateur radio license for nearly 30 more years, the seed had been planted and was waiting to germinate.

Of course, like most everyone else in the U.S., I was exposed to depictions of shortwave radio / ham radio in TV shows like Hogan's Heroes and in movies such as Independence Day, Frequency, and Terminator III. Those shows and films probably helped maintain a spark of interest for me over the years.

I can't guarantee you that by learning about and using ham radio you'll be able to save the world from an invasion by extraterrestrial beings or sentient machines or stop a serial killer, but I can guarantee you an otherworldly experience when you first connect with someone halfway around the world using a small high-frequency transceiver and a length of wire for an antenna, or when you hear through your handheld radio another operator read back your FCC-assigned call sign through a signal bounced from a tiny artificial satellite or the Earth's moon.

3

WHY HAM RADIO IN THE AGE OF THE INTERNET?

Like a lot of people these days, when the idea of pursuing an amateur radio license popped in my head, my first question was, Do people still do ham radio? Didn't that die with the birth of the Internet?

To find out, the first thing I did was run a search online to check out the status of ham radio. What I found stunned me. Not only is there a vibrant amateur radio community, but also the number of licensees is growing robustly, with more than 755,000 licensed in the U.S. alone, and about three million worldwide. In 2018, the number of new licenses granted in this country was over 30,000! There is a really informative online video explaining why ham radio still exists, entitled "Amateur Radio (Ham Radio) as Fast As Possible," which can be found at *https://youtu.be/ysOq6ywTSzU*. (The video actually is a disguised advertisement for a VPN app, but the first four minutes is devoted to the wonders of ham radio.)

There are enormous differences between the way people communicate through the Internet and how they do so over the airwaves. However, some of what we consider the Internet travels via radio signals and some of what we consider ham radio now takes place through the wired world. Like a rainforest, this new and beautiful world of electronic communications possesses tremendous "biodiversity."

For most people these days, using the Internet consists primarily of social media interactions with our "friends," searching for information, online shopping, and sending e-mail messages to specific people. Users tend to know who or what they're looking for. In contrast, using ham radio is more akin to fishing: you throw a line in the water, and you're never quite sure what (or who) you're going to pull up. Sometimes you get skunked, but much of the time you are handsomely rewarded by either reaching across the country or across the globe to connect with someone new, or at least you get to hear others doing so, often in exotic, faraway lands. For those who have established friends in the ham world, using their radios is a way for them to sit around a virtual fire and share stories. There is great camaraderie in the world of amateur radio.

Personally, I find ham radio addicting. I'm downright obsessed with it. I'm not sure I can effectively explain my fascination to someone who isn't inclined toward working the airwaves. My burning interest is some combination of a fascination with the technology and the underlying physics, a love of learning this multifaceted art and science, a quest for better preparedness in the event of an emergency, the competitive aspects of contesting and achieving awards, and a big dose of "intermittent reinforcement" from making contacts around the world.

For anyone who's ever played Pokémon Go on their smartphone, making QSOs (contacts) with a ham radio, especially DX (distant/foreign) contacts, has an interesting similarity to capturing an elusive and rare Pokémon. Just because you spot it, doesn't mean you'll catch it.

4

WHAT YOU CAN DO IN HAM RADIO

The fundamental purpose of ham radio is communication. Whether it is for socializing, disaster relief preparedness / disaster assistance, or an excuse to fiddle with electronics, ham radio operators are engaged in communicating with others.

Ham radio means very different things to different people. For some, the thrill comes from fast-paced **contesting**: trying to make as many contacts as possible in a fixed amount of time. For others, it's all about **DX**: making contacts with people in other countries— sometimes to have real dialogue, but often primarily to rack up a list of countries contacted and to receive awards. Some hams just want to be able to chat with people ("**ragchewing**"), wherever they may be. Others are "**preppers**" — those who actively prepare for all types of emergencies from natural disasters to civil unrest (or even the coming zombie apocalypse). Still others are in it for the joy of **building their own transceivers** or trying to connect with people with very low power (**QRP**). Others like to use **vintage equipment**. Some pursue communications relayed through **satellites** passing overhead, which can involve artificial ham radio satellites or bouncing signals off of the moon (**"EME" – Earth-Moon-Earth communication**)! Related to this, opportunities exist for **communicating with astronauts** on the International Space Station. Ham operators can send images through the airwaves: **slow-scan TV** (SSTV) on high frequency (HF) bands (still pictures), and **fast scan television** (FSTV) — moving video — on VHF/UHF bands.

Using different types of equipment, ham radio operators can talk to people locally (sometimes with VHF/UHF handheld transceivers called "HTs" – derived from Motorola's "Handie-Talkie" trademark), or communicate with hams over long distances, hundreds or thousands of miles, with HF (high frequency) transceivers.

One of the first activities many hams get involved in are **nets** – regularly scheduled on-air meetings of hams who have common interests. Many local ham radio clubs operate nets on VHF or UHF repeaters that anyone with a Technician class license can participate in, as long as they are close enough to reach a repeater, often done with a handheld transceiver. Nets also occur on the HF bands, with participants scattered across vast areas. Generally, a net control station (NCS) initiates the net operations and directs the net activities. Stations that want to participate in the net check in at the direction of the NCS. If you're not a regular net member, wait until the NCS calls for visitor check-ins. Give your call sign to check in.

Some of the topics mentioned above (e.g., contesting) are explored in greater depth in this book. Other subjects will be left for your continued exploration elsewhere. The purpose of this book is to help get you on the air as quickly and painlessly as possible so that you have the opportunity to explore various modes of operating. It would be more than a little difficult to run out of things to do in the fascinating world of amateur radio.

5

GET LICENSED QUICKLY

If you're someone who anxiously wants to take and pass one or more ham radio tests, perhaps you skipped right over the introduction in this book. As I stated there, let me give you a key tip:

> *Don't make the mistake of getting bogged down in trying to learn every aspect of ham radio before you allow yourself to enjoy actually doing it!*

Flight instructors invariably tell their students that a pilot's license is a license to learn. The same holds true for a ham radio license. Get it as quickly as you can, and then learn by doing.

Nevertheless, you are going to have to learn a certain amount of esoteric information at least long enough to answer a sufficient number of test questions correctly. It's well worth the effort.

Good news: Ham radio tests probably are not like the tests you remember from school. With ham radio tests, just like airplane pilot tests, the question and answer bank is completely available for you to review in advance. The purpose of the testing is not to hide the information you need to succeed. Rather, the tests push you to learn and provide the information to do so, so that you can operate properly and safely.

In the past, there were many levels of amateur radio licensure. Now there are only three classes of licensure: Technician, General, and Amateur Extra.

The intro-level license is Technician Class. The Technician test is relatively easy for most people. There are only 35 questions to answer and, according to the ARRL as of 2018, the pass rate is 81%.

Unfortunately, the Technician Class license is limiting in terms of operating privileges. As of the time of this writing, a Technician can only use HF (high frequency) for CW (Morse code), not for voice ("phone") communications (except for a sliver of the 10-meter band), which is usually done in single sideband (SSB) mode. High frequency SSB is where most of the ham radio action is, because it allows people to *talk* across the country and around the globe. CW will cover equal, and probably greater, distances, but not with the ease and richness of the spoken word. There is a proposal for Technicians to have new phone and image privileges on 80 meters (3.900-4.000 MHz), 40 meters (7.225-7.300 MHz), and 15 meters (21.350-21.450 MHz), but it has been kicking around for a couple of years, and I do not know if it will be approved.

On the other hand, a Technician can communicate fairly long distances using either satellites or terrestrial repeater networks to relay VHF/UHF signals. Also, Technicians have phone privileges on 6 meters, called the "magic band" for the long-distance propagation that sometimes occurs there. In addition, Technicians can use fast-scan television (FSTV). Those are pretty cool privileges.

Nevertheless, because the Technician Class license has limited HF phone privileges, I strongly recommend that you begin studying for the General Class license as soon as you are confident that you can pass the Technician test. Even if you do not think you are ready, take both tests on the same day, which usually can be done for a single fee. You might surprise yourself and pass a second test!

Securing General Class privileges will give you access to SSB voice communications on all ham radio bands, which is a big deal for most hams. (On the other hand, if you only need a ham radio license for the ability to engage in communications within, say, a 5- to 20-mile radius, the Technician Class license may be just fine.)

You can even take the Amateur Extra Class test in the same test session as the Technician and General tests. If you feel comfortable with the study material, then go for it! I did, and I was thrilled to pass all three tests on the same day.

There are two benefits to holding an Amateur Extra Class license: (1) you have access to the full spectrum of ham radio frequencies; and (2) you have the opportunity to obtain a coveted "1x2" or "2x1" vanity call sign. "N9CD" is a 1x2 call: a prefix of one letter, a number, and a suffix of two letters. When you are repeating your call sign thousands of times on the radio, you'll be happy to have a short call—like the difference between the names "Tim" and "Timothy." (You will find more about vanity call signs in Chapter 7, section H.)

So, here's the secret to passing the multiple choice ham radio tests: Study the test questions and the *correct* answers only. (A lot of test prep materials include all of the wrong answers along with the correct answers, which can lead to confusion on test day if you happen to recognize answers that are wrong.)

Following are the stats on the FCC question pools online as of 2020:

Question Pool	Effective Dates	Quest. Pool Size	Quest. on Test	Min. to Pass
Technician Class	7/01/2018 - 6/30/2022	423	35	26
General Class	7/01/2019 - 6/30/2023	454	35	26
Extra Class	7/01/2020 – 6/30/2024	622	50	37

A.   Test Prep

To begin the process of preparing for the ham tests, you need to review the test material. The last section of this book consists of a study guide for the Technician class license, including the entire pool of 423 questions, listing the correct answer only rather than displaying of the multiple choice options. This book provides you with all the information you need to pass that test.

When I was preparing for the tests, I used eBooks for all three license classifications, which I read using the Kindle app on my iPhone. Personally, I used the book series, *Pass Your Amateur Radio [Technician / General / Amateur Extra] Class Test – The Easy Way*, by Craig Buck (K4IA).

One day, I was not a licensed amateur radio operator. The next day, I achieved "Amateur Extra." I moved through the Technician and General Class material, and was doing great with online practice tests. For the Amateur Extra test, I never even bothered to read the narrative part of any book. Instead, I jumped straight to the test questions and answers. I took all three tests on the same day and passed them all. In the limited time I had, I could not have done this without focusing solely on the test questions and correct answers. That was plenty to learn but nothing extraneous to bog me down before I could get licensed.

There are a number of test prep books out there, including *The ARRL Ham Radio License Manual* (a large-format book with 288 pages in the 4th edition) and *ARRL's Tech Q&A 7th Edition*. Some books will provide you with a lot more detail to learn about ham radio than simply focusing on the test questions and answers. If you feel you have the time to read additional material before you wade into DOING ham radio, then, by all means, do so. I just know from my personal circumstances and experience that I wanted to dive into being a ham radio operator as quickly as possible, and that led me to focus on the test questions and correct answers only.

Although I did not use the site myself, other hams have told me that Ham Study (*hamstudy.org*) is an excellent site for test preparation. It is like other practice test sites, but a person can study all or only sections of the test. The site tracks one's progress and re-asks questions that a user has gotten wrong more often. The site has a button for a user to obtain an explanation for why a particular answer is the correct one. A disadvantage (in my mind) is that the site presents all of the multiple-choice test responses, including the incorrect responses, unlike this book. Of course, any website that allows a person to take practice tests (see the next section) is going to display the wrong answers with the correct ones. But I recommend that you do not use interactive practice tests as study material. Rather, use the tests merely to determine when the study material has stuck in your brain well enough to pass the test(s).

B.   Practice Tests

One you have studied the test material, you should take practice tests to determine how well you have retained the material and to decide when you are ready to attend a test session. Fortunately, there are websites with free access to practice tests using the actual questions on the exam, which are randomly sorted. The online tests will give you an immediate test score, and at least some will track how many questions in the test pool you have encountered as you take the test multiple times.

For my own test prep, I used the practice tests on the QRZ.com website: *www.qrz.com/hamtest*. You have to register an account on the site, but it is completely free. After you get your license, I can guarantee you will want a QRZ.com account anyway, as it is THE go-to site for looking up contact information, QSL preferences, and biographical information and photos of hams you encounter on the airwaves. The site also provides a free electronic logbook. (Logbooks are described elsewhere in this book.)

Ham Study (*hamstudy.org*) is another site for taking practice tests. Some of its features are described in the section above. To study on the go, use of a mobile browser on a smartphone or tablet always has been one way to access Ham Study. More recently, Ham Study developed offline-capable apps for most mobile and desktop platforms for users who wish to use HamStudy.org without an Internet connection.

Yet another option is the *aa9pw.com* site, created by ham Simon Twigger. Options include online practice for all three US written exams, email exams, java applet-based exams, and Morse code practice. Twigger also offers HamMorse, an iPhone app that allows one to practice Morse code on an iPhone or iPod.

The American Radio Relay League (ARRL) also offers an online resource that allows users to take randomly generated practice exams using questions from the actual FCC examination question pool. ARRL Exam Review for Ham Radio™ is free, and users do not need to be ARRL members. The only requirement is that users must first set up a site login (a different and separate login from one's ARRL website user registration). Visit *www.arrl.org/exam-practice*.

C.   Where/How to Take the Tests

Ham radio license tests are administered by Volunteer Exam Coordinators (VECs) who are approved by the FCC to administer exams. In the pre-COVID-19 pandemic era, most prospective hams would contact a local ham radio club to arrange for taking one or more of the amateur radio tests, usually for a $15 fee (which covers as many of the three tests that you may want to try on a single day). ARRL has a "Find an Amateur Radio License Exam in Your Area" search feature at *www.arrl.org/find-an-amateur-radio-license-exam-session*. It may be just as easy to use Google, enter your community name and "ham radio club" or "ham radio testing" to find a VEC nearby.

In-person tests remain an option, but online tests are available, too. A remote exam session is conducted using an online videoconferencing platform and a web-based examination system with on-screen tests. An advantage of such tests is the fact that such exam sessions are not restricted by location. Thus, an exam candidate is not limited to testing only with a team in his or her own state. Visit *www.arrl.org/FindOnlineExam* for more information.

On exam day, you will need to bring one legal photo ID or, if no photo ID is available, two forms of identification. (Requirements for minors are somewhat different.) In advance of the test, you will need to obtain a Federal Registration Number (FRN) (see *www.fcc.gov/wireless/support/universal-licensing-system-uls-resources/getting-fcc-registration-number-frn*). New license applicants should create an FCC user account and register in the FCC Commission Registration System (CORES). If you already have an amateur radio license, you will need to present a copy. Bring a couple of number two pencils with functional erasers and a pen. You are allowed an electronic calculator, as long as it does not have any formulas stored in memory. Don't forget to bring the examination fee (which you may need to pay in cash for in-person tests). The FCC now also requires test takers to supply an e-mail address.

A tremendous amount of resources exist to help you prepare for the ham radio tests. The testing cost is miniscule and the benefits of being licensed are great. With no CW (Morse code) test anymore, nothing stands in your way. You can do this!

6

GET ON THE AIR QUICKLY, PAINLESSLY, AND INEXPENSIVELY

A. Locate Resources to Help Through Rough Patches

Inevitably, you will hit snags on your ham radio journey. Even if somehow you manage to avoid that, you can always achieve success faster with help and guidance, rather than having to learn everything the hard way. Don't reinvent the wheel or the radio. This book may be an initial step for you to figure out how to get on the air quickly, but do not ignore other resources.

1. YouTube is a tremendous resource.

There are numerous YouTube channels promoting and supporting the ham radio community. Here is a short list of some of the many channels providing ham radio information:

- Ham Radio Crash Course
- Ham Radio 2.0
- Jim W6LG
- David Casler - Ham Radio Answers

Simply go to YouTube and search for these channel names to find a plethora of videos, or run a search in YouTube for a specific topic and see the video offerings from many content providers. If you goal is to watch videos on a specific subject, it is best to run your search directly in YouTube rather than in a general search engine such as Google.

2. There are articles online about every imaginable ham radio issue.

Run an online search for any ham radio question you may have, and you will find information galore. A couple online sources of information include:

- American Radio Relay League - ARRL.org
- License preparation and education service - HamRadioSchool.com

3. Join a local ham radio club and the ARRL.

I recommend that every U.S. ham join the American Radio Relay League (ARRL), the national association for amateur radio in the US. The organization publishes two magazines for members: *On The Air* for newer hams and *QST* for more-established operators. ARRL members get discounts on emergency communications training courses and have access to a variety of other resources to help develop and maintain the skills one needs to effectively perform public service communications. The organization also advocates for meaningful access to radio spectrum.

Don't stop at joining the national organization, though. ARRL appears to have more than 2,700 affiliated clubs. You can run a search at *www.arrl.org/find-a-club*. Alternatively, use a web browser and search using the name of your community and "amateur radio club." Clubs usually have monthly meetings, distribute newsletters to their members, and host events throughout the year. (In the age of the COVID-19 pandemic, many clubs have transitioned to using videoconferencing for club meetings. Now you can participate in meetings from the comfort of your home.) Club members are tremendous resources for new hams (see section on "Elmers" below). Many local clubs administer amateur radio tests, as well. Also, many clubs offer a free membership to newly-licensed hams for the first year.

4.   Find one or more "Elmers" to provide assistance and advice.

"Elmer" is a term in amateur radio for a mentor, a ham operator who provides encouragement, advice, and assistance to fellow hams. Most hams find an Elmer to help them get started. Given the complexity of taking principles from the ham tests and applying them to the practical aspects of getting on the air, having an Elmer is invaluable.

Unless you already have a friend or relative who is a ham, you probably will need to join a local club in order to meet potential Elmers. Do not be shy in asking for help. Every ham radio operator has experienced some struggles and many experienced guys seem willing to provide assistance and to share their love of ham radio.

I consider myself incredibly fortunate to have multiple Elmers. One of them gave me my first experience of seeing a ham radio contest in action (shortly before I was licensed, I got to watch him operate—I loved seeing the fast pace of the contest). He also gave me a spare G5RV (dipole antenna) that I still use as my one and only HF antenna, some books and magazines, and other miscellaneous bits of gear. Another Elmer helped me convert the G5RV into a ZS6BKW antenna (optimized for multi-band use), created a balun for the antenna (a device to help prevent common mode current from flowing on the antenna feed line), and trained me how to use the local ham club's HF radio through a remote connection (an Internet interface for controlling and operating a radio). Yet another Elmer provided advice about sending and receiving QSL cards with hams in far-flung countries.

B.   Equipment to Get on the Air

1.   Listen with webSDR

When I started considering the idea of becoming a ham (the second time around in 2019, not the first time I considered it back around 1990), I wanted to have the opportunity to listen in on what hams actually did on the radio. I briefly considered buying a shortwave radio receiver. However, I realized that once I got licensed, I would need a transceiver for transmitting *and* receiving, and a simple receiver would be unnecessary at that point. I figured it made more sense to save my money for an HF rig.

Fortunately, there is a way to eavesdrop on ham radio operators for free using nothing more than a computer and an Internet connection. WebSDR is a Software-Defined Radio receiver connected to the Internet, allowing many listeners to tune it and listen simultaneously. Visit *websdr.org* for links to receivers around the world.

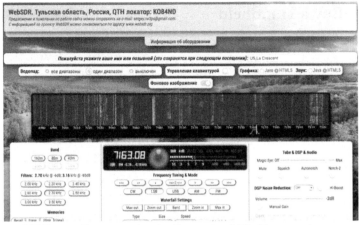

A webSDR site in Russia tuned to the 40-meter band. The blue rectangle across the middle is the waterfall display, and the vertical purple lines are radio signals.

The interface for webSDR receivers makes them easy to use. The ones I have seen all have a waterfall display, showing radio signals across whatever band the radio is tuned to. A user can click on a spot where a signal appears to tune the radio to that frequency, or type a particular frequency in an input field provided for that. Usually, there will be buttons to select different bands and to make other adjustments such as the mode (single sideband, CW, FM, etc.)

Even if you have an HF radio at home, webSDR allows a person to hear signals in other parts of the world at times when propagation conditions or one's own antenna limitations do not allow reception over the air. Also, if you do not have a waterfall display on your own rig, you can use a webSDR receiver to provide a waterfall display and to quickly jump through the bands to see where the action is.

You actually can hear hams talking in real time with webSDR. Once when I tuned in to a European webSDR receiver, I heard a particular operator from Slovenia who I have talked to multiple times. Then I turned on my HF transceiver and tuned to the same frequency. Sure enough, I heard the same guy "calling CQ" through my radio. I could hear him better on webSDR, though, because that receiver was located at least a couple thousand miles closer to him than I was.

2.   Buy a Handheld

Prospective hams quickly discover that the world of amateur radio is divided into two fairly distinct frequency realms: HF (High Frequency) and VHF/UHF (Very High and Ultra High Frequency).

High frequency communications are what enable long-range (around-the-world type) communications, whereas VHF and UHF generally are limited to shorter-range, line-of-sight communications, due to the way the signals do or don't propagate through the atmosphere. HF signals easily can bounce off the ionosphere and skip to other parts of the globe, but VHF/UHF signals generally pass through the ionosphere out into space (unless sporadic E propagation or tropospheric ducting kick in).

With an HF radio and a simple, inexpensive wire antenna, a ham can easily communicate with people across the country or across an ocean. It's truly amazing.

Although the lesser availability of atmospheric skipping of signals for VHF/UHF might seem like a big limitation on communications, it does have some distinct advantages. For one, VHF and UHF signals are needed to penetrate the ionosphere to communicate with satellites in orbit. Fortunately for hams, there are dozens of amateur radio satellites that can be utilized for free to take even a low-power signal from an inexpensive handheld transceiver and rebroadcast it hundreds of miles away.

For folks who aren't interested in tracking satellites and scrambling to make contacts during a brief pass from above, there are repeaters in place around the earth that will retransmit a signal from a VHF or UHF ham radio to greatly extend one's reach. Some repeaters are linked with others to extend the range even further.

For accessing repeaters on towers or satellites, a ham can do this with a base station, a mobile radio in a vehicle, or a handheld transceiver (called an "HT" for handy-talkie). Hams who live out in the boonies far from a club repeater may not find an HT particularly useful. On the other hand, it is possible to inexpensively construct a handheld beam antenna or rig up another type of antenna outdoors to greatly extend the range to reach more distant repeaters.

Cheap vs. Expensive HT Debate

Handheld radios can be purchased on amazon.com for as little as $25 or as much as several hundred dollars for higher quality versions.

When you start investigating HT options, you will quickly learn that a debate is raging within the ham radio community about whether hams should avoid (or buy) cheap radios from China (Baofeng/BTech or Wouxun) or spend maybe five or ten times as much on a better radio, generally made in Japan by Yaesu, Icom, Kenwood, or Alinco. As examples of this debate, there is a YouTube video entitled, "Why People HATE The BaoFeng Ham Radio!" and an online article entitled, "Should I Buy That Radio From China?"

There are several factors to consider when selecting an HT: price, features, and quality.

Price: On the basis of price, it seems like a no-brainer to buy a Chinese-made radio. How can you beat a dual-band (VHF/UHF) radio that only costs $25?! As of the time of this writing, a Baofeng UV-5R can be purchased on Amazon for $24.99 new. Amazing. In contrast, the Yaesu FT-60R costs $179.95 on Amazon, a whopping seven times more expensive! For the price of the Yaesu, you could buy a half-dozen Baofeng HTs and stash them in each of the family's cars, keep one in the house, another in the garage, and have one at the office, etc. A lot of hams do employ such a strategy, even if they also end up buying one or more Japanese HTs.

I happen to own a BTech UV-5X3 tri-band (2-meter, 1.25-meter, and 70-cm) radio, which I believe cost me about $65. I never use the middle band, and I could have saved myself some money by getting a dual-band radio instead. However, I had read that the tri-band model had better, newer firmware and functioned better, and I thought the third band would make the radio more versatile. (It turns out almost nobody uses the 1.25-meter band, at least in the U.S.)

You may wonder why anyone would spend money on a Japanese handheld, given how relatively pricey they are. The reasons come down to the next two factors.

Quality: Although I currently do not own a Japanese HT, I have read and heard from others that these have a much higher build quality than the cheap Chinese radios, which seems obvious given the dramatic price difference.

Another quality issue with handhelds, in general, is the "rubber ducky" antenna they come equipped with, which many hams describe as being little better than a "dummy load" (a device used for testing purposes, simulating an antenna to allow the transmitter to be adjusted and tested without radiating radio waves).

There is an issue not only with hardware quality but also with signal quality and reliability of cheap transceivers. With my BTech UV-5X3, I sometimes notice erratic functionality. For example, there are times when I am listening in scanning mode, and it starts to include the weather frequencies in the scan, which are programmed into the radio to be skipped.

As a more serious issue, one online article points out that amateur radio in the U.S. is regulated by FCC's Part 97 rules, including transmitter technical specifications in Subpart D, and that spurious emissions need to be well below the mean power of the fundamental frequency of the transmitter. Reportedly, the Baofeng UV-5RA dual-band transceiver was run through a battery of tests, and measurements showed that some or all of the Baofeng's harmonics were well above the FCC limits. However, there are quite a few comments posted to that article that suggest that at 5 watts of transmit power, the spurious signals are not likely to have much real-world impact.

One of my "Elmers," a very experienced ham, says:

> Put me in the camp that is firmly against cheap Chinese radios! ARRL tested several over the years and many, perhaps most, do NOT meet FCC spectral purity requirements. Even with a 5w portable, these things can put out spurs that could interfere with public service (police, fire, airport) communications.

That is a strong argument against buying a cheap radio. On the other hand, I rarely transmit with my cheap HT, mostly using it as a scanner and having it available in case of an emergency. If you're going to keep a radio in the glove compartment of your vehicle for emergencies or use it as a scanner, the device's spectral purity doesn't matter nearly as much. In any event, I don't believe anyone disputes that the Japanese HT's are of a higher quality all around than the Chinese radios.

Features: In addition to considering price and quality, the features offered on a particular transceiver affect the purchasing decision. Cheap Baofeng radios are relatively slow scanners. And although they can display two frequencies at once, they cannot actually use both at the same time. "Dual Watch" tries to emulate simultaneous use of both tuners. Yaesu radios, in contrast, are very fast at scanning and have dual VFO functionality that can receive from two channels at the same time. The Baofeng UV-5X3 has 128 memory channels, whereas a Yaesu FT-60 has a thousand. Also, the expensive models may come with a "spectrum analyzer" mode. In addition, pricier units might be equipped with Automatic Packet Reporting System (APRS), giving ham radio operators a way to transmit their location, allowing them to view the locations of other ham users, and send messages. There are many additional features that distinguish various transceiver models.

Another big factor for a lot of hams is the menu interface for programming the radio on the fly. Baofeng radios are notoriously difficult to program manually, whereas Yaesu radios have a reputation for being much more intuitive and faster to program manually. (More on programming radios below.)

If you merely want a handheld (or several) to occasionally connect with other hams through a repeater and/or to have available for emergencies, a cheap Chinese transceiver might be fine. But if you want a radio of higher quality and with more features, you will have to pay more for that.

Programming a Handheld

Acquiring a handheld VHF/UHF radio can be inexpensive and easy, but programming them can be quite intimidating. Unlike FRS (Family Radio Service) devices with just a handful of dedicated channels, ham radio HTs do not come out of the box ready to use, particularly because most ham radio bands are not channelized. If you did take one out of the box, charged the battery, and then started transmitting, you might end up breaking the law or even interfering with law enforcement, emergency, or military communications.

Ham HTs must be programmed to the local frequencies you need. A typical HT might have between 128 and a thousand memory channels. It will be your job to figure out what frequencies to program into those channels. Generally, hams include a number of local repeater frequencies, local law enforcement and emergency frequencies (both for routine monitoring, like one would do with a police scanner, and to have those frequencies handy in case of a real emergency), Family Radio Service (FRS) and General Mobile Radio Service (GMRS) frequencies (although the radio may not be legal to transmit on those frequencies due to exceeding power limitations, etc.), and channels for the National Weather Service.

PROGRAMMING FILE CHANNEL LIST
FRS - GMRS - PMR - MURS - BUSINESS
MARINE - WEATHER - SAR - HAM - VHF / UHF

FILE NAME: FRS_GMRS_PMR_MURS_BUS_MARINE_WX_HAM_2013F.CSV   INFO: RADIOFREEQ.WORDPRESS.COM

MEM CH SLOT	UHF VHF	CHANNEL DESCRIPTION	CHANNEL DISPLAY NAME	FREQUENCY RECEIVE	FREQUENCY TRANSMIT	OFF SET MHZ	PL	TONE HZ	MODE
0	UHF	FRS & GMRS CH 1	FRS 01	462.562500	SIMPLEX	0.0	TX PL	67.0	NFM
1	UHF	FRS & GMRS CH 1	FRS 1	462.562500	SIMPLEX	0.0	TX PL	67.0	NFM
2	UHF	FRS & GMRS CH 2	FRS 2	462.587500	SIMPLEX	0.0	TX PL	67.0	NFM
3	UHF	FRS & GMRS CH 3	FRS 3	462.612500	SIMPLEX	0.0	TX PL	67.0	NFM
4	UHF	FRS & GMRS CH 4	FRS 4	462.637500	SIMPLEX	0.0	TX PL	67.0	NFM

List of radio frequencies found online (excerpt).

Fortunately, there are helpful websites that provide current information on frequencies available by state and county, such as *RadioReference.com* and *RepeaterBook.com*. There even are computer programs like CHIRP that will download selected frequencies from those websites for directly programming into your radio.

To program dozens or even hundreds of frequencies into your handheld transceiver, you are not going to want to do that manually, as that would likely take hours, if not days, of tedious work. Instead, you probably will use the CHIRP program, a free, open-source tool for programming your amateur radio. It supports a large number of manufacturers and models, as well as provides a way to interface with multiple data sources and formats. Start by visiting *chirp.danplanet.com*.

Loc	Frequency	Name	Tone Mode	Tone	ToneSql	DTCS Code	DTCS Rx Code	DTCS Pol	Cross Mode	Duplex	Offset	Mode	Power
0	0.000000		(None)									FM	
1	146.970000	WR6ARC	Tone	131.8						-	0.600000	FM	High
2	146.640000	W0NE	Tone	100.0						-	0.600000	FM	High
3	146.835000	W0NE	Tone	131.8						-	0.600000	FM	High
4	147.150000	WD0HAD	(None)							+	0.600000	FM	High
5	147.285000	N0PDD	Tone	100.0						+	0.600000	FM	High
6	147.015000	N0ZCD	Tone	110.9						+	0.600000	FM	High
7	147.090000	N9ETD	Tone	131.8						+	0.600000	FM	High

Frequency list loaded into CHIRP (cropped image).

You will need a programming cable to connect your radio to a computer, and that is where a lot of problems arise, as there are knock-off cables that refuse to interface with some computers. When I bought my first HT, I repeatedly failed in my efforts to program it, because the computer could not see that the radio was connected, which apparently was due to the cable. Eventually, I contacted the president of the local ham radio club for help, and he loaned me a cable that did work properly. (Use the FTDI programming cable, which is available on Amazon.)

To do the programming, you can find a lot of guidance online, including written instructions, such at the CHIRP Programming Software Guide (*miklor.com/COM/UV_CHIRP.php#FAQS*) or the BaoFeng Radio - Chirp Software Programming Guide (*iwillprepare.com/ham_radio.htm*). In addition, there are helpful YouTube videos, including "How to Program a Baofeng HAM Radio with Chirp – TheSmokinApe" (*youtu.be/0l_kdktZAkI*), "How To Program a Baofeng Ham Radio Easy and FAST With CHIRP" (*youtu.be/MEpBo5lixsw*), and "How to Program the Yaesu FT-60R with Chirp" (*youtu.be/1uQcJ4g0akM*).

CHIRP allows you to save a copy of all of the radio programming settings on your computer. Once your HT is programmed, you can tweak the list later using CHIRP or manually change individual frequencies out in the field.

Once programmed, a handheld ham radio is an incredibly powerful tool in a very small package. It will allow you to talk to other hams locally, either directly in simplex mode (within a few miles) or farther through repeaters, or much farther away by using satellites passing overhead. It can serve as a weather radio, a receiver of FM broadcast stations, a scanner, an emergency communications device, and it may even include an LED flashlight!

3.   Use EchoLink®

Chances are, before you purchase your very first piece of ham radio gear, you already own sufficient hardware to communicate with other hams around the world. As long as you have an amateur radio license, you can use the EchoLink service, which can be accessed with a smartphone or a computer over the Internet using streaming-audio technology, or via a ham radio over the airwaves by connecting with an EchoLink-enabled repeater. The program allows worldwide connections to be made between stations, or from computer to station. According to EchoLink, there are more than 200,000 validated users worldwide — in 151 of the world's 193 nations — with about 6,000 online at any given time. The system has a validation process to ensure that only licensed hams participate. Visit *echolink.org* for more information.

4. Get Access to a Remote HF Rig

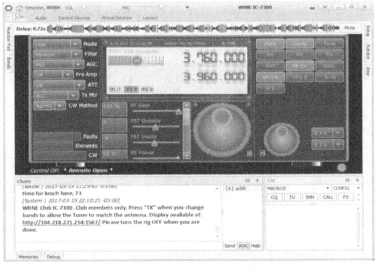

A ham radio owned by a club and remotely accessible through the Internet.

A remote HF rig, although accessed through the Internet, is not the same thing as webSDR receiver (a web-based Software Defined Radio) that one might use solely for *listening* to other hams. A remote rig is a computer interface for taking control of a transceiver to both receive and transmit. It is analogous to a remote control for a television, allowing a single user to change channels without physically touching the TV.

When I passed the ham tests, I could hardly wait to get on HF. Thus, I began searching online in earnest for used equipment.

However, I knew it was going to take a while to find a good deal on equipment, figure out what to do for an antenna, get an antenna mounted in the trees outside my house, and run a feed line into the house.

Fortunately, a ham I connected with through one of three local amateur radio clubs informed me that by joining a certain club, I could gain remote access to the club's well-maintained Icom 7300 through the Internet, using a program from *remotehams.com*. The same day I received confirmation of my license, I sent a letter with a check to the club to join, and I asked for information to connect with the remote rig. As a result, within two days of getting my license, I succeeded in getting on the air and logging three QSOs with hams in Utah, West Virginia, and Florida. I was thrilled!

The author's initial ham shack consisted of two computers and a handheld transceiver.

Even if your local ham radio club does not have a remote rig, there are a bunch of them out there on remotehams.com, and you can request permission to use some of them.

Assuming you are able to get access to a remote rig, for some people that may be sufficient as a long-term or even permanent solution, and they may not want the expense of obtaining their own rig and antenna, or they may live someplace where it is impractical to install a decent antenna.

5.   Look for Used Equipment

For me, I retained a burning desire to have my own HF radio, despite having access to an excellent remote rig. I figured that the process of learning how to acquire and assemble the various components for a ham radio operation is key to becoming a real ham. After all, it's all about having the knowledge to get on the air, right? Also, in some of my early QSOs I noticed that some hams seemed to be disdainful of connecting with a person on a remote rig, one of them telling me he does not log such contacts.

Because I was so excited to get my own rig, after I passed the ham tests I picked out and ordered a used rig even before my license was issued by the FCC (which I got notice of online ten days after taking the tests). Despite ordering a radio right away, my earliest QSOs were made with the remote rig, because it took another two or three weeks to get my rig, prepare an antenna, get an antenna up in the trees next to my house, and run the feed line into the house.

The author's IC-706MKIIG with MFJ-939 antenna tuner and Powerwerx power supply. It's a pretty basic shack, but a lot of QSOs have been made with it.

I focused my radio search on *eBay.com*, but there are other places to look for used equipment. Such sources include *QTH.com*, *eham.com*, and *QRZ.com*.

Many choices for a ham radio exist, and it can be incredibly confusing knowing where to start. What bands will you want to use? Do you want a portable rig? Are you going to be using CW or SSB? Are you interested in QRP (low power) operation? How big is your budget? Do you want the latest technology for ease of using digital modes and having fancy panadapters with waterfall displays to graphically show radio signals across a band? Do you ultimately plan to have more than one rig for different purposes, or do you want the "Swiss Army Knife" of ham radios (referred to as a "DC to daylight" rig) that will do it all? Those are all important considerations.

There are reasonably-priced *new* radios, too. The IC-7300 is a very popular transceiver that sometimes can be purchased for under $1,000, and it comes with a built-in antenna tuner and some fancy features.

6.   Build or Buy an Inexpensive Antenna

I think one of the toughest issues for a new ham to grapple with is deciding what antenna or antennas to install. There are a bewildering array of antenna types, including single-band, dual-band, multi-band, dipole, fan dipole, trap dipole, vertical, trap vertical, wire, random wire, loop, screwdriver, Yagi, J-pole, Slim-Jim, VHF/UHF and HF/VHF mobile. Some new hams get so bogged down with antenna decisions that it keeps them from getting on the air. Throw up a simple antenna to start. You'll have the rest of your life to perfect your antenna situation.

What you end up installing will depend upon factors such as what bands you want to operate on, how much space you have, whether you live in a neighborhood with restrictive covenants, and how much money you have to spend. Resonance and bandwidth are two properties for antennas that affect their performance relative to the frequencies you may want to operate on. The length of an antenna is a big factor affecting performance.

VHF/UHF Antennas

For VHF/UHF, you might simply use the rubber ducky antenna that came with you handheld transceiver (HT). However, those antennas are notoriously inefficient. An excellent substitute is a J-pole or a dual-band Slim Jim attached to the HT with a short length of coax. There are a lot of plans on the Web for building your own antenna, including from items as simple and inexpensive as some coat hangers and a connector. If you're handy with a soldering iron and are not uncomfortable with the idea of installing a UHF male PL-259 connector on the end of the coax, then by all means build your own.

Such antennas can be purchased inexpensively, too. Personally, I bought a dual band 2m / 70cm Slim Jim antenna with 10 feet of coax installed from N9TAX Labs (*n9taxlabs.com*) for a mere $25.99. Actually, I bought two: one with the 10-foot feed line attached for connecting to my HT and another, even cheaper one without the length of feed line that I connected to my base station with a much longer piece of coax that I bought separately.

HF Antennas

A ham with unlimited resources and space likely would install one or more beam antennas (such as a Yagi) on one or more towers. With such an antenna, an operator can experience significant gain, with the transmitted signal concentrated in a particular direction.

However, erecting a tower and installing a large beam antenna is expensive (perhaps $5,000 or more), requires a lot of space, and may draw the ire of neighbors and/or violate restrictive covenants. A full-size Yagi for 20 meters is about 32 to 35 feet wide. I've seen plans for a "low profile" 20-meter beam antenna, but even that is about 16 feet wide. If you want an antenna for the 40-meter band, plan on doubling the size, about 65 to 75 feet wide with a nine-foot-long boom for two elements, or a 40-meter coil-loaded Yagi at about 43 feet. One online article describes the size as "wider than the wingspan of a Piper Cherokee airplane."

For most new hams who lack the space, the money, or both for a big beam antenna, there are excellent and inexpensive alternatives.

I am far from an expert on antennas, but I know what "inexpensive" is, and I know what I have been able to do with a bit of wire. If you have a yard with a tree or two or three (and even if you don't have any trees), a wire antenna of some sort can get you on the air quickly and inexpensively.

A few antenna types to consider are single-band dipole, fan dipole, random-length dipole, G5RV, and ZS6BKW (a modified G5RV). I have seen single-band dipole antennas for sale (new) for under $40, and G5RV multi-band antennas for as little as $60.

Single-band dipoles are as simple an antenna as you could hope to build. If you are content to stick to a single band (e.g., 20 or 40 meters), such a dipole will be optimized for the particular band you select. If you have a big enough yard, you could set up multiple single-band dipoles, each with its own feed line, to work multiple bands. Alternatively, make a fan dipole by attaching multiple wires cut for different bands to the same center support and feed line.

Wire Dipole Antenna for Use on HF Bands

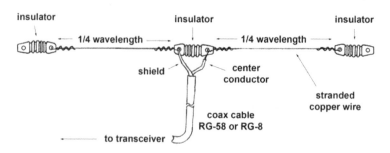

You simply need some copper wire, three insulators, and some coax feed line. Make each side of the dipole one-quarter of a wavelength long, with an extra half-foot on each end for trimming to tune the antenna. You can calculate the total length of both halves of the dipole using the following formula: 468/frequency (in MHz) = Length (in feet). You can find plans for building this online. Then hang it to span a couple of trees, or raise it up on one tree and stake down the ends in an inverted vee. For more information about simple wire antennas, visit *www.arrl.org/single-band-dipoles*. Also, ARRL has an article available for download from the June 1983 issue of *QST* magazine entitled "Wire Antennas for the Beginner" that explains how to construct end-fed random-wire antennas usable on all HF bands as well as various center-fed dipole antennas. Just search for the article title online.

For multiband use, you could buy or build a G5RV, a ZS6BKW, a trap dipole, parallel (fan) dipole, an end-fed random-wire antenna, or a random-length multiband dipole fed with ladder line (a/k/a window line).

The G5RV antenna is a dipole with ladder line feed line. The dipole elements are 51 feet each (102 feet total) and the impedance-matching feed line can be either 300 ohm (29 feet) or 450 ohm (34.0 feet). With an antenna tuner, you can use the antenna for multiple bands.

The ZS6BKW is an optimized variant of the G5RV, being 92 feet wide and 40 feet high, using ladder line for the vertical feed line. Because the feed line is "balanced" ladder line, which then connects to unbalanced coax to get to your radio, it is

Homemade balun to connect between coax & ladder line

necessary to include a balun (an electrical device to help prevent common mode current from flowing on the antenna feed line) between the feed line and the coax leading into your shack.

Unlike the G5RV, which requires an antenna tuner on all bands, the ZS6BKW is widely touted as capable of operating on 10, 12, 17, 20 & 40 meter bands, generally without an antenna tuner, although I use a tuner with my ZS6BKW. One ham posted in an online forum, "A ZS6BKW is NOT a no-tuner antenna. The best one can do is trim for no-tuner operation on a couple of bands, e.g. 40m and 20m," and I tend to agree with that. My ZS6BKW seems to work great on 40 and 17 meters, reasonably well on 20 meters, and is rather mediocre on 80 meters (although I have not spent much time on that band). Tuning definitely is important on 80 meters with my setup.

You may be wondering what an antenna tuner (a/k/a antenna tuning unit or ATU) is. It's a device connected between a radio transmitter and its antenna that matches the load impedance of the radio to the combined input impedance of the feed line and the antenna, thus lowering the standing wave ratio (SWR) to improve the effectiveness of the antenna. I use an MFJ-939I with my Icom transceiver and multi-band ZS6BKW antenna, and it cost about $150. You can see a photo of the tuner next to my IC-706MKIIG in section 5 of this chapter.

Don't Be Afraid to Run Coax into the House

One way or another, you will need to run a feed line from your antenna into the house, unless you plan to operate outdoors. Coax and ladder line can be routed through window frames or under sash windows or brought in through a hole in the wall or foundation. A reasonably simple solution for apartment dwellers is to use some 1/8" Teflon coax (such as RG-316) about 1 foot in length to pass through the window gaskets. The coax is lossy, but a short length won't affect the signals.

When I started out, I was intimidated by the idea of running a coax feed line from an external antenna through a wall into the house. I worried about hitting wires running through walls or creating a pathway for vermin to enter. I also had a silly notion that cutting a hole in an outside wall would lower the value of my house. One of my Elmers (a ham radio mentor) told me to be bold and happy that I was getting a chance to cut a hole through the wall and that I should embrace the opportunity as a rite of passage.

Nevertheless, I thought I figured out a way to avoid cutting a hole and having to run coax through the inside of the house. My house had already been wired for cable TV, and the basement home office that I planned to use as a ham shack had a connection for a wall-mounted TV. Coax ran outside to the supports for the deck where the Dish Network or DirecTV dish antenna had been mounted. I had taken that dish down, because that service had been replaced by fiber optic cable for TV and Internet, thus freeing up that particular coax.

This seemed like a great plan until I learned that coax used for cable TV does not have the right specs to use as feed line for a high frequency antenna. My heart sank when I realized that I actually did need to cut a hole in the side of my house.

It occurred to me that the best entry point into the basement was the area around the electrical and gas service entrance into the furnace room. Pipes and wires were already run through the wall, and the ceiling in the furnace room was open, allowing me to see the exterior sheathing just above the foundation. At the service entrance on the outside, there was a square plastic framework for a dryer vent that had never been used. That seemed to be the perfect spot to create an opening.

My first step was to look at where this vent-opening-to-be was relative to the wires and pipes entering the house, and to measure the distance away from those reference points. Then I moved to the interior of the furnace room to determine where the hole would pop out.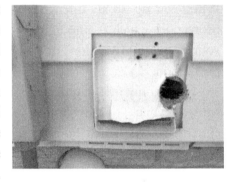

Feeling confident no electric wires or gas pipes would be in the way, I drilled a small (1/8-inch?) hole to see exactly where I would pop through. With that hole and drilling a couple additional holes, I was able to refine the location. (The hole looks off center, but that could not be avoided due to metal on the inside of the wall.) Once I was confident that I had located the center for a hole, I used a hole-cutter to create a 1.5-inch hole. (I wanted the hole roomy enough to comfortably fit at least two feed lines through it, including passing the PL-259 connectors through.)

In some instances, it would be advisable to insert a short length of PVC pipe into the hole, run the cable(s) through that, and then pack some sort of putty, caulking or other sealant around the cable(s). With my house, at the point of entry into the furnace room there was just a single layer of plywood (the exterior sheathing) with no internal drywall or whatever to pass through. Thus, I felt a length of pipe would be pointless.

Once the feed line was in the house, I pulled it across the open ceiling of the furnace room and supported it with some existing brackets designed for holding up water pipes. From there, I snaked it over the suspended ceiling in the basement and then popped it out into the room with the radio (the "shack"). It ended up being a relatively easy process.

Lightning Protection

When you rig up an antenna, do not forget to implement lightning protection, which is important for protecting the safety of you and your family, protection of your electronic equipment, and fire prevention.

I set up my first antenna in early winter when thunderstorms are extremely rare, which bought me a few months to add lightning protection, but I recommend including it when you first install the antenna. You can start with an in-line coax lightning arrestor tied into a

grounding system outside, before the feed line enters the house. I use a lightning static / surge protector rated for DC to 1000 MHz and 400 Watts with SO-239 connectors, which I bought online for about $25. Some are significantly more expensive.

During thunderstorm season, also consider disconnecting your ham radio gear from feed lines and unplugging the equipment from outlets when not in use.

There is a lot more to this subject, and I recommend checking out resources available through ARRL (*www.arrl.org/lightning-protection*) and talking to your local ham radio club to find out who can share their expertise with you.

Hopefully, now you won't be intimidated by the myriad antenna options. Make or buy some form of wire antenna, throw it up in the trees, hook up some lightning protection, run some coax into the shack, add an antenna tuner (especially if you plan to use a multi-band antenna), and start connecting with other hams!

7

OPERATING PRACTICES

A.   Ham Etiquette

Following proper etiquette on the air is important for amateur radio to work smoothly and to avoid "road rage," since hams generally have to police themselves.

When a ham operator fails to use good etiquette, it can result in frustration for others, and makes everyone else's experience less enjoyable. For example, it is rude and annoying when people tune-up (operate their antenna tuner) on a frequency on which people are already communicating, as it can blast a loud tone into people's ears and disrupt QSOs in progress. Find a quiet frequency a few kHz away to tune the antenna before responding to a ham calling CQ.

It may be hard to appreciate how important good etiquette is until you join a pileup of dozens of hams all trying to contact the same DX or special event station all at once.

As one example, in an unruly pileup, stations start calling the DX operator before he or she has finished his or her current QSO. As a result, the DX station or the other party to the QSO may have to "shush the crowd" and repeat the exchange of information or closing remarks, thus delaying others from making contact. It is not unusual to hear a DX station reminding those calling to wait for the DX operator to invite the next round of calls by saying "Q-R-Zed." Some DX and special event operators allow bad behavior to persist, usually resulting in the pileup getting more an more unruly and chaotic.

Some DX operators are particularly sensitive to certain bad behaviors, and they are proactive in trying to keep others from engaging in those behaviors by posting "rules of engagement" on their QRZ profile pages. For example, I know of a DX operator in Central America who has such rules posted, and he informs callers that if you violate his rules, such as calling while he's in the middle of a QSO, he will make note of the offending call sign and refuse to make contact with that person. This is a powerful incentive to behave, especially if you need that particular country to advance toward the DXCC Award (an award for making contacts in 100 different countries).

DX Code of Conduct

Reviewing the DX Code of Conduct is a good place to start for understanding etiquette on the air, as it also applies to making other contacts:

- I will listen, and listen, and then listen again before calling.
- I will only call, if I can copy the DX station properly.
- I will not trust the DX cluster and will be sure of the DX station's call sign before calling.
- I will not interfere with the DX station nor anyone calling and will never tune up on the DX frequency or in the QSX slot.
- I will wait for the DX station to end a contact before I call.
- I will always send my full call sign.
- I will call and then listen for a reasonable interval. I will not call continuously.

- I will not transmit when the DX operator calls another call sign, not mine.
- I will not transmit when the DX operator queries a call sign not like mine.
- I will not transmit when the DX station requests geographic areas other than mine.
- When the DX operator calls me, I will not repeat my call sign unless I think he has copied it incorrectly.
- I will be thankful if and when I do make a contact.
- I will respect my fellow hams and conduct myself so as to earn their respect.

Another good way to develop a feel for good operating practices is to spend time listening to other operators. It does not take long before you start to recognize bad behavior on the air.

Legal Rules

In addition to voluntary rules of etiquette, there are some legal rules for how to conduct oneself on the air. Prohibited transmissions are governed by 47 CFR §97.113. Generally, no amateur station shall transmit communications for hire or for material compensation, music, broadcasting / one-way communications, communications intended to facilitate a criminal act; messages encoded for the purpose of obscuring their meaning; obscene or indecent words or language; or false or deceptive messages, signals or identification.

Terminology

To communicate effectively on the air, it is important to learn basic terminology, including the most common "Q signals." Q signals are a set of three-letter codes, all starting with the letter "Q" that serve as shorthand for commonly-expressed information. Q signals were developed for commercial radiotelegraphy and carried over to amateur radio. In addition to saving time (especially for hams using Morse code), they serve as a common language for ham operators around the world.

Q signals (Q-code) are for CW (Morse code) operation. Generally, Q signal use for voice communications is improper. Use plain English instead. Nevertheless, even non-CW hams need to know a few Q signals lest they be confused when other hams use them in phone mode. A few Q signals are quite handy even for voice communication, such as giving a signal report like "5-5 with QSB" for a perfectly readable transmission with fairly good (but fading) signal strength. (Of course, one could say "5-5 but fading" as an alternative.)

I have seen a list of as many as 65 Q signals, but you do not need to memorize more than a few (at least if you restrict yourself to voice communications). Here is my "top 10" list – the ones you will hear regularly on the air:

QRM (often stated as "Q-R-Mary" to distinguish it from QRN or "Q-R-Nancy") – a reference to interference from another station, usually someone who is not maintaining appropriate frequency spacing (3 kHz for SSB).

QRN "Q-R-Nancy") – interference from static.

QRP Decrease power. I never hear this as a request or a question like the other Q signals, but only as an adjective to describe whether someone is operating at low power (e.g., "I'm using my QRP rig.")

QRT Stop sending. Usually used by a ham at the end of a QSO when that ham is going to leave the air, e.g., hanging it up for the night and going to bed.

QRZ (Almost always stated as "Q-R-Zed") It means, "Who is calling me?" and is used as an invitation to other hams to initiate a QSO. For example, a DX station who has a pileup of hams calling will, after finishing a QSO with one ham, give his or her call sign followed by "QRZ" to invite the next round of people calling.

QSB   (Sometimes stated as "Q-S-Baker") It means, "Your signals are fading." Sometimes a station will be really strong one moment and hard to hear the next. When giving that station with a lot of variability a signal report, you might say, "Your signal is 5-7 with a lot of QSB."

QRL?   Is the frequency in use? (Used in CW only. With voice, one would always say, "Is the frequency in use?")

QSL   "I am acknowledging receipt" or, as a question, "Do you acknowledge receipt?" This one drives me a little nuts, because some hams overuse it to a severe degree. Some guys will say, "QSL, QSL…" at the beginning of every transmission, as if to say, "Yeah, I agree with you." This Q signal is intended to be used to confirm a transmission, such as when one ham gives a signal report: "You are 5-9 in Wisconsin. QSL? The ham receiving the signal report can acknowledge receipt of the report simply by saying, "QSL" or "QSL the 5-9 in Wisconsin" to show the other ham that the information truly was received.

QSL is used in another context: confirming a contact for purposes of preserving a memory of a contact (especially the joy of connecting with someone in a distant country, known as "DX") and/or for awards. This type of QSL is accomplished through the exchange of QSL cards or electronically through online services such as Logbook of The World (LoTW), QRZ.com, or eQSL.com (addressed elsewhere in this book).

QSO   An over-the-air contact or conversation.

QSY   Change to another frequency. If you find that a QSO on a nearby frequency is bleeding into yours, you may need to QSY. Also, hams using "calling frequencies" to establish connections need to QSY to free up the calling frequency for others.

QTH   A ham's location. My QTH is La Crescent, Minnesota, for
       example.

In addition to learning how to use and interpret the most common Q
Signals, there are other bits of terminology helpful to learn. For
example, when a ham operator wants to get on a frequency and start
calling for people to respond for a QSO, it is customary to start with
"CQ, CQ, CQ ...." (More on that below.) To politely end a QSO, say
"73" (an old telegraph code that means "best regards"). A man on the
radio sometimes is referred to as "old man" ("OM" in CW) regardless
of age, a woman is referred to as "Young Lady" ("YL"), and a wife as
"XYL."

Unless you want to be looked down upon and maybe even
ridiculed, avoid using Citizens Band (CB) terminology on ham radio.
For example, hams do not acknowledge a communication with "10-
4." Instead, they are likely to say "QSL" or "Roger, roger." (See "QSL"
in the list of Q Signals above.)

B.   Calling CQ

When you get your license, you may be reluctant to make contacts
(QSOs) on the radio. There are several reasons for such reluctance,
and one of the prime issues is being unsure of how to properly initiate
a contact, and being afraid of sounding stupid or inexperienced.

Knowing the correct vernacular to use on the radio and employing
scripts can go a long way toward improving one's comfort level talking
on the radio.

When a ham wants to initiate an on-air contact, they call "CQ"
(literally "seek you" – calling any station). There's a procedure for
doing that, which I will explain below. However, I think it's easier to
start out answering other hams calling CQ than calling CQ oneself.
That way, you don't have to worry about whether the frequency is

clear, because the other ham has taken responsibility for that. Also, you know that other ham already wants to speak to someone; otherwise, that person would not be calling CQ. In addition, you will get multiple opportunities to hear and write down their call sign before you answer, making sure you know who you will be talking to.

I recommend having a computer in front of you with the QRZ.com website up. When you hear a call sign, type it into the search box on the webpage and hit enter. This provides a good check **In the decades before QRZ.com existed, hams had to spend $40 a year for the US and DX "Callbook" — the ham radio equivalent of a telephone book — which became obsolete before each edition was printed.** that you heard the call sign correctly, as the vast majority of hams are included—even operators in other parts of the world. If you entered a valid call, a ham's biographical information should appear, and you will get a name, address, and other information about the person that can be useful for generating ideas for conversation.

I made the mistake of not using QRZ.com when I started on HF. On my fourth QSO, just a few days after receiving my license, I talked to a guy named Paul in Arizona, but I somehow wrote Paul's call sign down wrong. Because I didn't check it online until later, I didn't realize it was wrong. (The call sign turned out to not be in the QRZ database, and I could not find it in the FCC database either.) As a result, I have not been able to properly log the contact, and I do not know how to reach him to get the correct information. (Paul, if you're reading this, our QSO was at night on 11/26/2019, and my call sign at the time was AE0HW. Shoot me an e-mail, please.)

This reminds me: get your biographical information on QRZ *before* you start operating. That way, you'll look more established, and other hams can look you up and maybe even send you a QSL card.

Make sure to tell the first ham you speak to that it's your first QSO. I did, and I was rewarded with a QSL card from a nice gentleman in southwestern Utah—a wonderful token of that first contact. (More about QRZ.com bios and QSL cards can be found elsewhere in this book.)

Whether you are having a QSO using SSB (voice) or CW (Morse Code), there's a certain exchange of information that tends to occur to make a properly loggable QSO. I suspect that CW tends to be a little more structured than SSB, because with voice the rate of communication is faster (150 WPM for normal speech), you'll glean more from listening to someone engaged in a previous QSO, and you will pick up subtle cues from the other person's voice, including unexpected accents for their QTH (home location), like the guy I talked to with a British accent but who is living in a western state. SSB QSOs can be fairly unstructured, except in contests where there is a specific exchange to be made. Let's start with CW, because it provides simple example of how the exchange of information might occur.

Simple CW QSO template

First listen for anyone else using the frequency. Then check if the frequency is in use before calling CQ. You don't want to step on someone else's conversation. You might not be able to hear one side of a QSO but the other person listening might be in your transmit range. In CW, send "QRL?" or "QRL DE [your call sign]?"

> CQ call / answer
> RST (signal report) / Name / QTH (location)
> Rig (transceiver) / Power (watts) / Antenna / WX (weather)
> Thanks / QSL info (e.g., direct, buro, LoTW) / 73

"RST" is the signal report. Hams can't hear their own signal, so they must rely on the person at the other end of the QSO to let them know how well they can be heard. With SSB, one exchanges the "RS" part: "readability" on a scale of 1 (unreadable) to 5 (perfectly readable); and "signal strength" on a scale of 1 (faint signals, barely perceptible) to 9 (extremely strong signals). With CW, the whole "RST" is given, with "T" being "tone" on a scale of 1 (very rough) to 9 (perfect tone with no trace of ripple or modulation).

Some hams seem to tend to hand out 59s to almost everyone, maybe so that they don't have to think about it. I have had many QSOs with people who have had trouble copying my call sign yet they've given me a 59. Note that in many contests in which contestants must exchange a signal report (such as in the ARRL International DX Contest), everyone, almost without exception, gives a 59 to avoid slowing down the process of making QSOs and to avoid mistakes in the logbook. If every QSO is a 59, then the only information you need to worry about copying correctly is the other party's call sign.

QTH, as noted above, is one's home location, unless one is operating elsewhere, like with Parks On the Air (POTA), Winter Field Day, or operating a mobile station.

Below is a sample script for calling CQ on SSB. Substitute your call sign for mine:

> CQ, CQ, CQ calling CQ [20/40/80] meters. This is N9CD calling. November 9 Charlie Delta, N9CD in southeast Minnesota calling CQ [20/40/80] meters. CQ, CQ, CQ [20/40/80] meters. This is N9CD calling. November 9 Charlie Delta, N9CD in southeast Minnesota calling CQ [20/40/80] meters and standing by for a call.

Not everyone who drives a car has to understand how things work under the hood, but it is essential that each and every driver, without exception, understand the rules of the road. Similarly, learning how to connect with others and conduct oneself on the air is equally if not more important than understanding details of the technology underlying ham radio communications.

C. Logging contacts

In the past, ham operators were required by law to log their on-air activities, and their log was subject to FCC inspection. While keeping a log is not a legal requirement now, it is an excellent idea anyway. For one, it provides a record of when you have been transmitting, in case a neighbor later complains about radio interference. If your log does not show you on the air when the complaint is made, it can help prove that it was not your station causing interference. (That's the boring reason for logging contacts.)

The primary reason most hams log contacts is because making contacts is what ham radio is all about. Your log provides a way to look back and see if you have spoken to a ham before and how long ago. It provides a way to verify how far the reach of your station is. Your list of contacts is a measurement of accomplishment, too—a point of pride for a well-built station. And for hams who pursue awards (e.g., "Worked All States," DX Century Club) or participate in contests (e.g., North American QSO Party, ARRL International DX Contest, state QSO parties, and the ARRL Sweepstakes), logging is a way to verify contacts and calculate points.

Please note that hams normally do *not* log contacts made on repeaters with VHF or UHF. Such contacts are not made due to the strength of your own rig, but are due to the power of the repeater to retransmit your signal. If you hold a Technician Class license and only use an HT (handheld) for local contacts, you will not need to bother with a log. On the other hand, if you use an HT to connect with hams hundreds of miles away through satellites, you may want to log those contacts. Or if you're among the rarer number of technicians who know CW, you can log CW contacts made on HF. Contacts made on HF generally are what hams log.

You may want to start with a paper log, especially if you are not logging a lot of contacts. You don't even need a formal logbook: a simple notepad will do, although ARRL has nice spiral-bound paper logbooks. When I started out, a ham in my local club gave me a paper logbook, which I used for a while before switching to a computer-based log.

While paper can work fine, most hams use electronic logbooks these days. There are several reasons for doing so. Electronic logbooks provide a mechanism for verifying QSOs for contests and awards.

They're also a convenient way to provide QSLs to other hams who may want confirmation of their contact with you, perhaps for awards they are seeking, or just for the pride of confirming contacts.

A lot of the ham radio software out there is for Windows computers. If you are a Mac or Linux user, your choices will be more limited, but software does exist.

Following are some examples of electronic logbooks. Some of these I have tried personally, but most of them I learned about from other hams.

N3FJP Amateur Radio Software

N3FJP sells "Amateur Contact Log" and a wide range of contest logging programs for the Windows environment. One can download and use the programs for a free trial period. According to the report of the results of the 2020 Minnesota QSO Party, about 30% of participants used this logging software, to give you an idea of how prevalent it is. I personally use this software and have been quite happy with it. In addition to using Amateur Contact Log as my day-to-day log, I have used half a dozen contest modules, such as for the ARRL International DX Contest and November Sweepstakes. It seems like a dated program, but it's rock-solid and easy to use. Amateur Contact

Log can be purchased individually for $24.99, or one can buy it and all of the contest logs as a software package for $49.99 (as of spring 2020). More information can be found at *www.n3fjp.com*.

DXLab

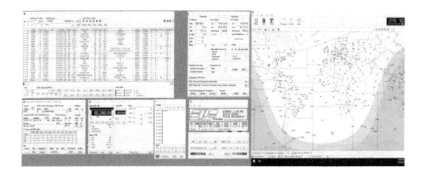

DXLab is a freeware suite of eight interoperating applications that can be installed independently in any order. When multiple applications are running, they sense each other's presence and automatically interoperate to support one's Amateur Radio DXing activities. One experienced ham told me that DXLab probably is the most popular of the newer logging programs. (Its low price [free] probably contributes to its popularity.) It can be a bit complex for the beginner to use. It can be used for contesting, but it is not a dedicated contesting log. Since it is free, it does not hurt to experiment with it. See *www.dxlabsuite.com* for more information.

Logger32

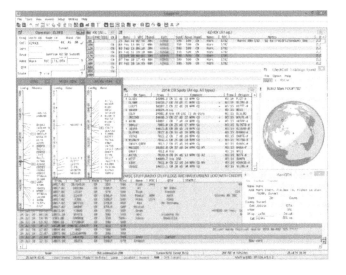

Logger32, a 32-bit Amateur Radio logging program currently supported to run under Windows 7, 8, 8.1 and 10. It has been developed to be a highly user-configurable general purpose amateur radio logbook with computer control support for many radios and antenna rotators. It is not a contesting log. The program is free!

www.logger32.net

Log4OM 2 (Log for the Old Man)

Log4OM is a FREE program that works with Microsoft Windows, starting from Windows 7. The website (*www.log4om.com*) lists a long range of features, but I have no experience with the program.

RUMlogNG

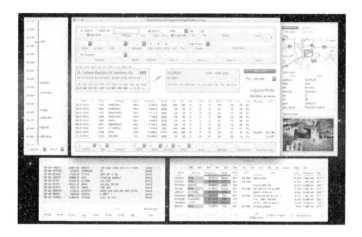

RUMlog Next Generation is a free Mac program created in Germany. I have a friend who uses the program, and he seems happy with it. According to the developer's website:

> RUMlogNG is a HAM radio logging, QSL handling and printing tool, especially made for the short wave DXer, made by a DXer. Basic logging features are included for the higher bands up to 1.2 cm and for satellite. RUMlogNG can handle an unlimited number of logs and an unlimited number of QSOs per log. Clublog data are used for automatic DXCC recognition. A contest module is included. Run RTTY directly using Fldigi or your Elecraft transceiver.

I have not personally tested the program, but free is a good price.

MacLogger DX

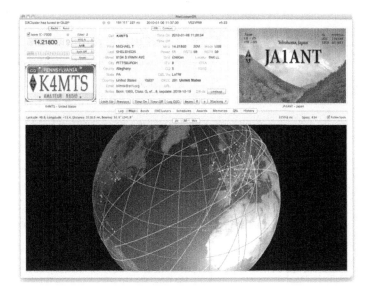

MacLogger touts itself as the "Premier Logger" for the Mac. Here's what the website has to say:

> MacLoggerDX supports more than a hundred radios, automatically tuning to the spots you are interested in, swinging your beam around. Alerting you to rare contacts or Band Openings and looking up, displaying on 2D, 3D and Satellite Maps and logging your contacts to a super fast sql database.

It sounds like a wonderful program. However, a single user license for the software costs a whopping $95.00. MacLoggerDX V6 must be registered to work beyond the 15-minute time limit. (You can run the 15-minute demo as many times as you like.)

If you are intent on using a Mac for logging, you might want to download both RUMlog and MacLogger and compare the two to see whether to spend the money on MacLogger or stick with the free program.

CQRLog

CQRLog promotes itself as the "world's best linux logging program." It provides radio control (support of 140+ radio types and models), DX cluster connection, online callbook, a grayliner, and internal QSL manager database support. CQRLOG is intended for daily general logging of HF CW and SSB contacts and is strongly focused on easy operation and maintenance.

Logbook of the World (LoTW)

Logbook of the World is a web-accessed database provided by the American Radio Relay League to implement contact verification among amateur radio operators. Using LoTW, radio amateurs can claim and verify contacts made with other amateurs, generally for claiming credit for operating awards, such as DX Century Club (DXCC). Thus, LoTW actually is a digital QSL service rather than a logbook *per se*. To use LoTW, one needs a separate digital logbook to periodically transfer digital files to LoTW to verify contacts with others who also upload their digital logbooks. (LoTW is discussed in greater detail in the *QSL Options* section elsewhere in this book.)

QRZ.com

QRZ.com provides a free online logbook. An advantage of this is that you can access your logbook on any web-connected computer or smartphone, regardless of operating system. A big disadvantage is that you do not have possession and control of your data. Also, I have heard stories of problems hams have had with incorrect data when they have tried to apply for the DXCC award (for making contacts with 100 different countries).

The QRZ logbook was my first digital logbook, although I have moved on to using N3FJP software as my primary log (see below). Even though it is no longer my primary log, periodically I export my log data from N3FJP and upload to QRZ. I just like having my QSOs accessible over the web, I welcome other hams to see my contacts (the most recent 15 are displayed on my QRZ profile page), and QRZ provides a way to confirm contacts with others using that logbook.

Logging contacts is fun for multiple reasons, including working toward awards. But the greatest satisfaction may be having a way to remember contacts with hundreds, and maybe even thousands, of new friends you will make on the air.

D.   Contesting

Ham radio contests are opportunities to make contacts with large numbers of ham radio operators in a short amount of time. Some contests run for only a few hours, while others go for 24 to 48 hours, usually on a weekend. Each contest has it's own particular goals, but all of them encourage ham radio operators to fire up their rigs and communicate with one another.

Contesting is a very different activity than "ragchewing." When contesting, ham operators exchange small, specific pieces of information to confirm the contacts for purposes of verifying points and prizes. For example, in the **North American QSO Party (NAQP)**, operators exchange only their call sign, first name, and state. For **Winter Field Day**, operators exchange their call sign and a short code with the "entry class" (the number of stations in your entry that are capable of simultaneous transmission, like if a ham club is operating two stations under the club call sign), the category (a letter identifier to specify whether the station is operating at home, indoors away from home, or outdoors), and the ARRL section identifier (a state or portion of a state, such as "WCFL" for west central Florida, or "DX"). For the **ARRL International DX Contest** and the **CQ World Wide WPX Contest**, participants exchange the sequential number of each QSO for themselves, along with a signal report (generally just "59" for SSB [voice] and, I presume, "599" for CW [Morse code]). For example, when I make my first contact in the DX contest, I would tell the other ham that my number is "001" or simply "one," whereas the other ham may be on QSO "057" or "1,024" or whatever. You enter the exchanged information in your log, and then you upload the log to the contest. Your logged information needs to match the information logged by the hams you connected with in order to be counted.

My first experience with contesting was as a spectator. After I had passed the ham tests but before my call sign was issued, I had the privilege of watching an experienced ham participate in the **ARRL Sweepstakes**. I enjoyed the fast-paced action of hundreds of ham operators all on the air at once, working stations up and down the bands. A year later, I got to operate in the Sweeps from my own shack.

My first participation in a contest happened by accident. I did not realize that the **North American QSO Party** was taking place on a particular weekend. I got on my radio to make contacts on 20 meters, and I quickly came across a "pileup" of ham operators trying to make contact with a domestic station, which is unusual outside of contests and some other activities like Parks On The Air (POTA). More typically, pileups occur when hams in the US have an opportunity to make contact with a station in another country—a "DX" contact. On this particular Saturday, I noticed that other hams were exchanging only call signs, first names, and state—no signal report, which is more typical. I connected with the guy, who happened to be in Massachusetts, and I followed the example of the other hams as to the information I exchanged. (Which is a good reason to listen before calling on the air.)

After that initial contact, I started hearing hams on other frequencies calling "CQ NAQP." I figured out that a contest was going on, and I searched online for a contest schedule to learn what the rules were. I ended up making 32 contacts that afternoon in about 2.5 hours of total operating time. That was way more contacts than I had ever made in a single day, which I think was a max of three or four per day previously. My logbook started to fill up quickly with people from around the country. From that moment, I was hooked on contesting!

I decided that I wanted to be proactive about contests—to plan for them and to know what needed to be done in advance. I discovered a handy online calendar of ham radio contests:

www.contestcalendar.com/contestcal.html

Another helpful resource is a bi-weekly e-mail from ARRL with a contest update. Getting on that e-mail list will help you be prepared for upcoming contests.

After NAQP, I only had to wait a week for the next contest, which happened to be **Winter Field Day**. However, I still didn't understand that Winter Field Day was a contest very much like NAQP, and I did not know the rules. I was under the impression that Winter Field Day was just about local ham clubs gathering and setting up radios outside in the cold and making a few contacts. I attended my local club's Winter Field Day event, in which they had a couple of radios set up in an enclosed park shelter and powered by a portable generator. After a couple hours hanging out there, I went home and turned on my ham radio. That's when I realized that a large number of hams were busy trying to make as many QSOs as possible. I jumped into it and managed to make about 22 contacts.

State QSO Parties: Each state has a QSO party designed to encourage contacts with every county in the state. These are a great way for new hams to experience contesting—not overwhelming like CQWW might be.

Since I was a Minnesota resident at the time I became a ham, I decided that I was going to make a strong effort in the MN QSO Party, which fell on February 1, 2020. The week of the contest, I purchased N3FJP's "Amateur Contact Log" and their whole collection of contest logging programs, including one for MNQP. With a dedicated logging program and an attitude that I was going to really go for it, I managed to log about 130 contacts, which was a much bigger number than the 20 or 30 contacts I had managed in earlier contests. I even managed to snag a certificate for "New Record Houston MN Single Operator Phone Category" in the 2020 Minnesota QSO Party. (I'm not sure if I had any competition in my county.) I had a tremendous amount of fun and connected with people from all around the country and a few foreign countries.

With the MNQP under my belt, I was ready to ratchet things up even further for the **ARRL International DX Contest** for SSB in March 2020. Fortunately, my wife was going to be out of town a big chunk of that weekend, which meant I could be undistracted for hours and hours.

But I had an even bigger prize fall into my lap. A ham friend of mine with a 500-watt amplifier and a beam antenna was going to be out of town, and he agreed to let me use his rig remotely through remotehams.com. With the tremendous power of his rig, I suddenly found myself slicing through huge pileups all over Europe and South America to have hams pick me up on my first, second or third call, rather than calling for 20 to 40 minutes to hope that I might make a single DX contact.

I didn't do anything earth shattering, but I blew away my earlier accomplishments, and I felt like I was on top of the world. Instead of picking up 130 or so *domestic* contacts, I managed to log about 230 QSOs with *DX* contacts in a single weekend, and I added about 45 countries to my total!

I realize that there are hams who log *thousands* of contacts in that contest, but I never said I was good at it. I'm guessing I could have logged another 50 or 100 contacts if my wife hadn't returned home by mid-day Sunday and if I didn't have some connection problems with the remote rig from time-to-time, which forced me to use a 100-watt rig with no beam antenna for stretches of time. In any event, I was ecstatic at the end of that weekend. I also managed to about double my total number of QSOs—in a single weekend!

Yet another fun contest is the **CQ World Wide WPX Contest** (yet another one that totally sneaked up on me). The WPX Contest is based on an award offered by CQ Magazine for working all prefixes. Held on the last weekend of March (SSB) and May (CW), the contest draws thousands of entries from around the world. I had not cleared by schedule for this and did not install the contest logging software. Also, I only had access to 100 watts and no beam antenna. In addition, I chose to avoid making many US contacts, preferring to focus on DX stations. As a result, I only managed to make about 30 contacts, but I did pick up a new state or two and two new countries, and I had a lot of fun.

The **North American SSB Sprint** takes a somewhat different approach that can be confusing to the uninitiated, especially for hams (like me) who don't bother to read all of the rules ahead of time. The Sprints are short, intense competitions lasting only four hours and using only the 80, 40 and 20-meter bands.

After having participated in several contests, I thought that all I needed to know was the "exchange" – the information that hams have to exchange and log to be considered a valid QSO in the contest. I was mistaken. I did not realize that the SSB Sprint has a special QSY (change frequency) rule: If any station solicits a call (e.g., by sending "CQ," "QRZ?"), that station is permitted to work only one station in response to that solicitation. The station must thereafter move at least 1 kHz before calling another station, or at least 5 kHz before soliciting other calls. Once a station is required to QSY, that station is not allowed to make another contact on the vacated frequency until or unless at least one subsequent contact is made on a new frequency.

Because I was unaware of the special QSY rule, I got confused during the first couple of contacts—losing track of who I was speaking to, since the Sprints' unique QSY requirements eliminate the usual approach in contests of dominant stations sitting on one frequency and running others. After a couple confusing exchanges ending with the other ham telling me, "The frequency is yours," I looked up the rules and figured out what was going on. Despite my initial confusion, I ended up making about 80 contacts in the four hours. I truly had a lot of fun!

Contests are a great way to learn how far your signals will reach, to quickly add states and countries to your log, and to test and improve your operating skills. If you like fast-paced action you definitely will want to try contesting. Even people who like to take things slow can dip their toes into contesting—it's not necessary to go all in. You can do as much or as little as you want to maximize the fun of it.

E. Chasing DX

"DX" stations are people operating in other countries or "DX entities." Many islands count as separate DX entities for purposes of the DXCC Award. For example, Guernsey is one of the Channel Islands in the English Channel near the French coast, it's a self-governing British Crown dependency, and it counts as a separate country for ham operators. Puerto Rico is a U.S. Territory, but it counts as DX. So do Alaska and Hawaii for those of us in the lower 48 states.

Chasing DX—the process of trying to make contacts with amateur radio operators in other countries—is one of the most rewarding activities for most hams. It is a way to test one's equipment and skills, as well as being a way to communicate with, and learn about, dozens or even hundreds of exotic locales.

As I mentioned earlier in this book, for anyone who's ever played Pokémon Go on their smartphone, making QSOs (contacts) with a ham radio, especially DX (distant/foreign) contacts, has a lot of similarity to capturing an elusive and rare Pokémon. Just because you spot it, doesn't mean you'll catch it.

To be effective in chasing DX, you need an understanding of where and when the action is on the various bands, since propagation, and operating times in various parts of the world, evolves throughout the day and night.

The DXMAPS "QSO/SWL real time QSOs" website (*www.dxmaps.com*) provides an excellent illustration of how propagation changes throughout the day, and it is one of a number of online services useful for spotting DX operators on the air at any given time.

During the day, 20 meters is great for reaching places like Europe, the Caribbean, and South America—especially on weekends. At night on 20 meters, I sometimes notice hams in New Zealand and Australia talking to hams in the U.S.

On 40 meters, I get good DX propagation at night. For example, from my QTH in Minnesota, throughout the winter I usually hear one or more hams in South Africa just about any day by tuning around the lower end of the Single Sideband portion of 40 meters (7.128-7.175), and I have made several contacts there. I've also made a variety of European contacts at night on 40, including Spain, Austria, and Bulgaria. In addition, I've picked up contacts with Hawaii on 40 meters at night.

I've noticed excellent propagation on 40 meters to Europe on winter Saturdays around sunset (say 5:30 pm CST), at which time it is 11:30 p.m. or 12:30 a.m. in much of Europe. The hams seem to be up and on the air then, because they don't have to get up early for work the next day.

On 80 meters at night, I have had QSOs with hams in Costa Rica, Alaska, and Serbia.

International contests are the best way to add a lot of DX contacts in a short amount of time. ARRL International DX Contest and the CQ World Wide WPX Contest are a couple prominent examples. In my first experience with the ARRL DX Contest, I managed to log about 230 QSOs and add about 45 countries to my total in a single weekend!

There are a lot more resources out there to learn about chasing DX. Although I haven't had a chance to read it yet, the most highly experienced DX'er I know recommends the book *The Complete DXer* by W9KNI. My friend states, "This is probably the best book written for learning about chasing DX. It's sort of a cross between a novel and a textbook. HIGHLY recommended."

Chasing DX is a ton of fun. If you are going to do it, be cognizant of the DX Code of Conduct, listed previously, to help ensure a smooth, enjoyable experience for all.

F.   Awards

Many hams pursue awards. Awards are issued by organizations such as the American Radio Relay League (ARRL) and others.

DX Century Club (DXCC) is Amateur Radio's premier award that hams can earn by confirming on-the-air contacts with 100 countries. Other awards include Worked All States (WAS), which is the most popular ARRL award. Worked All Continents (WAC) is the oldest award. Visit www.arrl.org/awards

Working toward awards is both very exciting (a process full of anticipation) and frustrating at times. It took me almost exactly a year from the time I got licensed to earn the Worked All States award (phone mode). (A person could do it much faster using digital modes or mixed modes.) For the DXCC Award, as of early 2021 I made phone contacts in 100 countries, with 93 countries confirmed either through Logbook of The World (LoTW) or through QSL cards.

I am waiting for confirmations from a mere seven DX operators. Some guys do not use LoTW, making it necessary to send QSL cards. Unfortunately, that process can be very slow. Some hams take many months to respond, some may not respond at all. For example, I had a QSO with a ham in Costa Rica in February 2020, but I did not receive a QSL card in response to mine until late October of that year. (I think the guy had confirmed the contact on LoTW much earlier.)

Also, some countries might not deliver mail from the U.S. For example, I have had both QSL cards I sent to Venezuela rejected with the notice "RETURN TO SENDER" on one side and "MAIL SERVICE SUSPENDED" printed on a label on the other side.

Although I do not use QSL bureaus to exchange QSL cards, I understand that one can wait *years* to get a card back that way, which seems crazy to me.

Thus far, I have made only one contact apiece with Denmark and Morocco, and I have made two contacts with Malta and Ukraine, for example, and I have yet to receive any form of confirmation from these hams. I have four contacts with Bonaire, but still no confirmation. To count these countries toward the DXCC Award, I either have to make more contacts in those countries, or I have to wait for these hams to eventually confirm. Personally, I find it hard to wait, especially since I otherwise would have achieved DXCC.

(For more information on the topic of confirmations, see the next section on QSL options.)

Other entities and websites besides ARRL, such as eQSL, offer awards, although ARRL awards seem to carry the most prestige.

Hams using the QRZ.com online logbook can qualify for awards on that website, including 50 Unites States Award (like WAS), DX 100 World Award (like DXCC), World Continents Award, Master of Radio Communication – Europe, and Master of Radio Communication – South America, among others.

I do not know what percentage of hams try for, or get, awards. Personally, I find it very motivating for making contacts. I hope that you find it a rewarding process, too.

G.   QSL Options

There have been a number of aspects of QSL'ing that I have found confusing since getting into ham radio, and it has taken some research, both online and from asking other hams, in order to understand the QSL process. Questions included: What is QSL? What are "green stamps?" What are IRCs? How do QSL bureaus work? Do I need QSL cards? How can I get on LoTW? Would it be helpful to be on eQSL?

The Q code "QSL" when followed by a question mark ("QSL?") means "Do you confirm receipt of my transmission?" "QSL" without a question mark means "I confirm receipt of your transmission."

During the early days of radio, the ability for a radio set to receive distant signals was a source of pride for many people. Listeners would mail "reception reports" to radio broadcasting stations in hopes of getting a written letter to officially verify they had heard a distant station. Eventually, as the volume of reception reports increased, stations began sending post cards containing a brief form that acknowledged reception. Collecting these cards was popular in the 1920s and 1930s, and reception reports were used by early broadcasters to gauge the effectiveness of their transmissions.

QSL Cards

Some DX QSL cards collected by the author.

Like in the old days with radio listeners, ham radio operators can exchange QSL cards to confirm two-way contacts. Such a card contains details about one or more contacts, the station and its operator. Generally, this includes the call sign of each station participating in the contact, the time (Coordinated Universal Time - UTC) and date when it occurred, the radio frequency or band used, the mode of transmission used (e.g., SSB, CW), and a signal report.

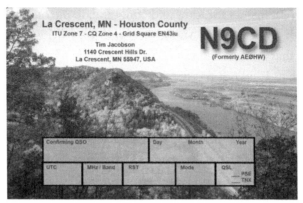

The author's QSL card, using a photo taken by the author as the background.

QSL cards have sentimental value as a way to remember a contact, possibly with someone in a distant land. They also have value for confirming contacts for purposes of awards, such as "Worked All States" and "DX Century Club" (DXCC) for working 100 different countries.

For decades, paper QSL cards were the only option for confirming ham radio contacts. However, now with the Internet a variety of QSL options exist, including ARRL's Logbook of the World (LoTW), QRZ.com, eQSL, and others, although the these methods are not equal in their effect. To prove contacts for prestigious ARRL awards, one needs to either use LoTW or QSL cards, or a combination of those two QSL methods.

When I got into ham radio, I wondered whether anyone used paper QSL cards anymore. They certainly *are* used, but I do not know what percentage of hams use them. One ham friend of mine who holds the DXCC award, tells me, "The volume has gone way down, but they are not yet extinct." They are especially popular in some countries like Germany and Japan. If you are going for an award and one or more of the contacts you need for an award does not use LoTW, you may be forced to use a traditional QSL card even if you otherwise don't want to mess with them.

The guy I had my very first QSO with after I got licensed was nice enough to send me a QSL card and congratulate me on my license. He was in Utah, and I was in Minnesota, and the card is a great way to remember that contact.

After that, I did not receive any QSL cards for about 2 months. It wasn't until I participated in a couple of contests that I started to receive more cards. At least one of those was from a ham who, according to a note he included with his card, "badly needed" confirmation of a contact with Houston County, Minnesota where I was transmitting from during the Minnesota QSO Party. In addition to sending me his own card confirming the contact from his side, he sent me a generic QSL return card, all filled out for me, so that I could easily confirm the contact. This way, even if I didn't have my own supply of QSL cards (which I didn't at that time), he could secure confirmation from me for an award. (I presume he is trying to get a "Worked All Counties" award.) Of course, he included a self-addressed, stamped envelope for me to return the card.

From Amateur Radio Station: N9CD							
Operator: Tim Jacobsen			Comments: 2020 MN				
Address: 1140 Crescent Hills Dr.			QSO Party				
City: La Crescent							
State: MN Zip: 55947							
County: Houston Grid:			73 es DX!				

CONFIRMING QSO WITH	DATE			UTC	MHZ	RST	2 WAY
	DAY	MONTH	YEAR				
	1	Feb.	2020	2234	7.287	59	SSB

QSL: ☐ PSE ☒ TNX

N9TT Generic QSL #9

Generic QSL return card

If you are going to use paper QSL cards, there are a variety of ways to get them printed. It is possible to print them at home, with the quality depending on what kind of computer printer and paper you have, as well as the level of your design skills. There is a website that provides a free "QSL Card Creator": *www.radioqth.net/qslcards*

You can have a local print shop/copy shop design and prints cards. In my case, since I know how to use Adobe Photoshop and InDesign, I designed my own cards and had a local copy shop print them. Getting 200 cards cost me just over $60.

Alternatively, there are plenty of printers that specialize in QSL printing. You can see a large number of card design examples on their websites. For example:

www.cheapqsls.com   *www.lz1jz.com*
www.kb3ifhqslcards.com   *www.ux5uoqsl.com*

Some of places, especially the cheaper ones, may not use custom artwork, whereas other printers will. Even if you don't have any design expertise, you may be able to supply a good digital photo to the print shop for them to incorporate into the design, whether it be a picture of your antenna, your ham shack, or local scenery.

QSL cards can be requested or sent in a few different ways (direct, QSL manager, QSL bureau, and OQRS (On-line QSL Request System)), and there are methods to help ensure you receive a reply.

QSL Direct

With this method, mail your card directly to the ham operator. It is the fastest method and provides the highest percentage of replies (99% according to one source).

Addresses generally can be found on the ham's QRZ.com profile page or, for U.S. hams, you also can look up addresses in the FCC's online Universal Licensing System "Amateur License Search" (*wireless2.fcc.gov/UlsApp/UlsSearch/searchAmateur.jsp*).

For securing a QSL card in reply, include a self-addressed envelope (SAE) with your QSL card. If sending to another ham in the U.S., make it a self-addressed, stamped envelope (SASE).

If you are sending the card and return envelope to a foreign country, obviously you can't simply use domestic postage on either the envelope you are sending out or the return envelope. For the envelope you have addressed *to* the foreign ham, you can purchase Global Forever international rate stamps from the U.S. Postal Service for $1.20 as of the date of this writing.

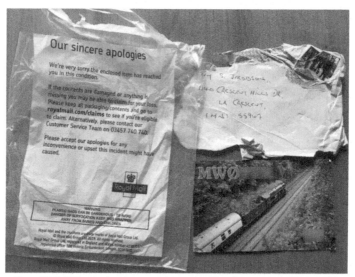

Some DX QSLs have a long, rough journey, like this one from Wales, UK. Fortunately, even though the envelope was mangled, the card was intact!

The trickier part is the return envelope. One common method of covering the postage cost for the replying ham is to include a couple U.S. dollar bills ("green stamps"). In the past one could use an International Reply Coupon (IRC) purchased at the post office. However, IRC's are no longer sold in the US or accepted in most countries. Green stamps are good; foreign postage is better, especially for casual ops (*www.airmailpostage.com*); many guys now accept PayPal, which some may think is even better. Look at the ham's QSL preferences listed on their QRZ.com profile, and follow those preferences.

QSL Managers

Many active DX stations use a QSL manager, especially when mail to the DX country is difficult. A ham's QRZ.com profile likely will identify any QSL manager. Information about the manager must be added to your outgoing card. Going though QSL managers requires you to pay postage both ways but results in excellent return rates and a fairly good turnaround time, in general. It definitely is faster than using a QSL bureau, but it can get expensive if you send out a lot of these.

QSL Bureaus

Sending QSLs via bureau (a/k/a "buro") involves a country's centralized amateur radio association QSL bureau, which collects and distributes cards for that country. On the positive side, this saves postage fees for the sender by sending several cards destined for a single country in one envelope, or large numbers of cards using parcel services. However, in exchange for lower postage, there is a delay in QSLs reaching their destination. Reportedly, it can take a year or even several years for a bureau to have enough cards to send to a particular destination. (I have not personally tried this yet.) In addition to such incoming bureaus, there are also outgoing bureaus in some countries. These bureaus offer a further postage cost savings by accepting cards destined for many different countries and repackaging them together into bundles that are sent to specific incoming bureaus in other countries. Most QSL bureaus operated by national amateur radio societies handle both incoming and outgoing QSLs (with the exception of the United States) and are coordinated by the International Amateur Radio Union (IARU).

An experienced DXer, W0VTT, tells me, "Buro is still useful for garden variety stuff. Many DL & JA [German and Japanese hams] still send buro cards, some do it for every stinking QSO."

For incoming cards, look at *www.zeroburo.org* and decide which option to take. You send them money, envelopes, labels, and whatever else your letter manager wants, and they send you cards every now & then. For outgoing QSLs, look at *www.arrl.org/outgoing-qsl-service* and follow the instructions. (One needs to be an ARRL member.) Many serious DXers with volumes of contacts still use the buro because it is so much cheaper than sending direct.

Club Log OQRS (On-line QSL Request System)

Club Log's OQRS offers an Online QSL form to request a QSL card, paying the other party for the postage cost via PayPal. Since the request is sent electronically, you do not have to wait for your QSL card to arrive via mail in a foreign country before the other party to the QSO responds. See *oqrs.net* and *clublog.org* for more information.

Personally, I avoided getting an account at clublog.org until I made contact with a station in Western Sahara that only QSLs via OQRS direct. Signing up turned out to be an easy process.

Club Log provides other features. When you register, you upload your digital logbook data file. Then you can use Club Log to generate a QSL chart, for example, displaying a list of countries you have made contact with and showing the status of those that have been confirmed (per QSLs entered in your logging program) and those contacts that match the log of another Club Log user. Club Log informed me that I have the opportunity to confirm contacts with 64 DX entities through its OQRS service. Fortunately, I had already confirmed most of these through LoTW (see below) and/or QSL cards. I did find a contact that I had been unable to reach via postal mail who is registered for OQRS. I have requested a card.

Logbook of the World (LoTW)

Logbook of the World is a web-accessed database provided by the American Radio Relay League to implement contact verification among amateur radio operators. It's  the gold standard for QSLs. Using LoTW, radio amateurs can verify contacts made with other amateurs, generally for claiming credit for operating awards, such as DX Century Club (DXCC). More than half of my QSOs have been confirmed through LoTW.

Even if you are not interested in pursuing awards yourself, I recommend getting registered for LoTW as a courtesy to other ham operators who will be looking for confirmations for their own pursuit of awards. You will make other hams very happy if you confirm your contacts with them, and who doesn't want to make hams happy?

To use LoTW, one needs a separate digital logbook to periodically transfer digital files to LoTW to verify contacts with others who also upload their digital logbooks.

Before QSOs can be submitted to LoTW, one must install the free TQSL application (Trusted QSL) on one's computer. TQSL will enable the user to obtain a Call Sign Certificate that identifies that person as the source of the QSOs one submits, and will also enable the ham to define a station location that specifies the geographical details of his or her operating location.

It is a bit of a pain to set up. After installing the software, there is a process whereby ARRL sends a postcard to the registered address of the licensee with a code to enter into the software in order to verify the identity of the person. (This is more secure than what I have to do for online banking, which seems rather extreme to me. It's not like there is money handed out when someone gets a DXCC award. Geez....) After that initial difficulty and delay, using LoTW is super easy. Usually, one merely clicks an "upload to LoTW" button in one's electronic log program, and the transfer happens automatically. Start the process of signing up for LoTW as soon as you get your call sign, so that you will be ready when you start making contacts.

QRZ.com

As mentioned in section C of this chapter, QRZ provide a free online logbook. It is another way to confirm contacts with others, but only for those using the QRZ logbook. When you log your contacts on QRZ, the site will automatically confirm contacts with other users who have logged their contacts there. A gold star will appear next to QRZ-confirmed contacts. You can, and probably should, use a different logging program as your primary log and periodically transfer standardized data files to QRZ.com to update it if you choose to keep a logbook there.

eQSL

Example of a customized eQSL card the author received from a French ham.

eQSL.cc bills itself as "the first and only global electronic QSL card exchange for amateur radio operators and SWLs [shortwave listeners]. It is designed to be the fastest, easiest, and cheapest way to exchange QSO confirmations, eliminating the cost and time that regular QSL cards have required for the past half century. With a larger membership than the entire ARRL, eQSL.cc is THE place where everyone exchanges QSLs quickly and easily. It has also become one of the largest awards organization [sic] for amateur radio with over 205,474 eAwards issued."

In contrast with these grandiose statements by eQSL, a very experienced DXer told me, "eQSL is worthless. Don't waste your time." The problem is that eQSL's awards don't seem to carry much weight. Nevertheless, I have an eQSL account. It's free, after all, unless you want to supply your own image for the eQSL card. It is quick and easy. As of the time of this writing, I have received about 200 eQSLs from other hams out of a total of 1,200+ QSOs, including numerous eQSLs from DX stations (Russia, Brazil, France, Italy, Slovakia, Guatemala, Spain, Guernsey, St. Martin, Aruba, Peru, Lithuania, Colombia, Chile, etc.). (In contrast, I have about 690 QSLs through LoTW.) Sometimes, an eQSL is the only confirmation I receive for a QSO in a particular country, and other times it may arrive weeks or months before I get any other confirmation.

I think it is fun to use eQSL, especially since one receives digital images of cards, unlike LoTW where all you get is a line of data for each QSL. Using eQSL certainly does not detract from any other method of confirming contacts. Since not everyone uses paper QSL cards or LoTW, it is one more method to confirm contacts that might not be important for ARRL awards or worth the postage, but still is nice to have documented.

From my perspective, QSLs, whether on paper or electronic, are one of the best parts of amateur radio, providing a nice remembrance, even serving as a trophy for valiant, persistent efforts to reach a ham in a distant land.

H.   Vanity Call signs

When the FCC issues a license and call sign to a new ham, there is an opportunity to apply for a vanity call sign (only *after* the initial call sign is issued). Like with a vanity license plate, you may be able to pick your own call sign (perhaps with your initials, for example), if it is available and if it is within certain parameters. When I first learned about vanity call signs, I dreamed up a whole bunch of them that sounded great, but were not available to me, because they represented countries other than my own or may not have been available to anyone. For example, I'm a pilot, and "P1LOT" seemed like a great call sign to have. But it's not a valid call sign. ☹

Call signs have a prefix (one or two letters), followed by a single numerical digit (0 to 9), followed by a suffix of one to three letters. In the U.S., call signs begin with A, K, N, or W.

Amateur Extras have the opportunity to obtain a coveted "1x2" or "2x1" vanity call sign. Short call signs generally are coveted for being able to say them (or transmit with CW) quickly on the air, especially during "pileups," and also to show that one is an Amateur Extra.

There are a number of resources online regarding vanity call signs. One good place to start for general information is *www.arrl.org/vanity-call-signs*.

I think the best online search feature for identifying available call signs is *www.radioqth.net/vanity/available*:

Display Options				
Format	1x2	Prefix	K	
District	3	Suffix		Search

Callsign	Available Date	Status
K3AV	02/28/2020	Available in 13 days
K3BE	07/17/2021	Available in 518 days
K3BP	04/28/2021	Available in 438 days
K3BT	01/08/2021	Available in 328 days
K3CB	01/27/2020	Application Filed
K3CP	09/14/2021	Available in 577 days
K3CV	04/27/2021	Available in 437 days

The site allows you to search under any of the available U.S. prefixes, any district (the numeral in the call sign), and to select 1x2 (e.g., N9CD), 2x1 (e.g., KB7A), 2x2 (e.g., AE0HW), 1x3 (e.g., W2AAA), or 2x3 (e.g., WA1ELS) configurations. The search feature will tell you if the call signs are already available, and if not, the date they will become available.

Important Note: You cannot apply for a call sign before the first day of its availability. If you try, your application will be rejected. Also, don't apply after the first day it becomes available. If someone else applies for the call sign on the day it becomes available and you wait until the next day, you will lose the opportunity to have it assigned to you.

Make a list of a number of call signs you might be interested in. You can apply for as many as 25 at one time, ranked in order of preference. If you get beat out of some of them, you still have a chance to secure a vanity call sign. I applied for nine at once—most of them 1x2 or 2x1 call signs—and I ended up with the third one on my list. Be aware that you might have to wait weeks, months, or even years for your favorite call sign to become available. If someone else is using a particular call sign, keeping moving and find another.

Get registered with the FCC's COmission REgistration System (CORES) in order to be able to apply for a vanity call sign: *https://apps.fcc.gov/cores/userLogin.do*. Do so as quickly as possible, because there can be delays.

While you are waiting to find out which call sign will be assigned to you (and you will wait—the process takes 18 days, for reasons that I'm not going to explain), there is a wonderful online tool, the Amateur Extra 7 Query tools, to determine how many people are competing for the same call signs and who is most likely to get a particular one: *www.ae7q.com/query*. Click the "Pending vanity application statistics" gray button. From there, you can see which ones are available to applications, which ones are in competition, etc. At the moment I am writing this for example, I can see that 12 people have applied for the 1x2 call sign W5CF.

As a ham, you're likely to say your call sign over the air thousands (if not tens or hundreds of thousands) of times, and you'll be making contacts with people around the country and around the world. Your call sign might become your nickname with your ham friends. You might as well pick a personalized or conveniently-short call sign, especially since the process is free (as of the time of this writing).

8

EXPLORING THE AMATEUR BANDS

The different frequency ranges available to amateur radio operators are not inherently better or worse than each other. Nevertheless, hams develop their own preferences, given various characteristics that can be exploited to achieve different goals.

High Frequency (HF) communications enable long-range (around-the-world type) communications due to the way the signals bounce off the ionosphere and skip to other parts of the globe. With an HF radio and an inexpensive wire antenna, a ham can easily communicate with people across the country or across an ocean.

In contrast, Very High Frequency (VHF) and Ultra High Frequency (UHF) generally are limited to shorter-range, line-of-sight communications, because the signals pass through the ionosphere out into space instead of reflecting back to earth. Although the general lack of atmospheric skipping of signals for VHF/UHF might seem like a big limitation on communications, it does have some distinct advantages. For one, VHF and UHF signals are needed to penetrate the ionosphere to communicate with satellites in orbit. Fortunately for hams, there are dozens of amateur radio satellites that can be utilized for free to take even a low-power signal from an inexpensive handheld radio and rebroadcast it hundreds of miles away, which is truly remarkable.

For folks who aren't interested in tracking satellites and scrambling to make contacts during a brief pass from above, there are repeaters in place around the earth that will retransmit a signal from a small handheld ham radio to greatly extend one's reach. For accessing repeaters on towers or on satellites, a ham can do this with a handheld transceiver (called an "HT" for handy-talkie). These little radios can be purchased on amazon.com for as little as $25 or as much as several hundred dollars for higher quality versions with more features.

Also, there is a difference between VHF/UHF and HF for sending images. Ham operators can send slow-scan TV (SSTV), which are still images, on high frequency (HF) bands. On VHF/UHF bands, hams also employ fast scan television (FSTV), which consists of video.

A.   VHF/UHF

The Very High Frequency (VHF) range is considered to run from 30 MHz up to 300 MHz. Ultra High Frequency (UHF) consists of 300 MHz to 3,000 MHz (3 GHz).

As noted above, VHF and UHF generally are limited to line-of-sight communications. On the other hand, sporadic E and tropospheric ducting can occur on 6 and 2 meters, greatly extending the range of signals. Also, meteor scatter is a daily phenomenon on 6 meters, which can be utilized for specialized digital communications.

The limitation of line-of-site communication is most often overcome through the use of amateur radio repeaters-- device that receives a weak ham radio signal and retransmits it at a higher power so that the signal can cover longer distances. Generally, repeaters are mounted on high towers, tall buildings, or hilltops to provide even greater range. Many repeaters are maintained by local ham radio clubs. Typically, repeaters can be found in the VHF 2-meter (144–148 MHz) and the UHF 70-centimeter band (420–450 MHz) (and rarely on the

1.25-meter (220–225 MHz) band). Repeaters operate in duplex mode, with a separate transmit and receive frequency.

Amateur operators tend to program their VHF/UHF radios to include a number of local repeater frequencies, simplex calling frequencies, local law enforcement and emergency frequencies for routine monitoring, like one would do with a police scanner, Family Radio Service (FRS) and General Mobile Radio Service (GMRS) frequencies (although the radio may not be legal to transmit on those frequencies due to exceeding power limitations, etc.), and channels for the National Weather Service. Websites providing current frequencies available by state and county include *RadioReference.com* and *RepeaterBook.com*. (A discussion of programming radios with the frequency offset, etc. is found in Chapter VI.)

As an alternative to using repeaters for VHF/UHF communications, hams can make direct (simplex) contacts (in which a single frequency is used for both transmit and receive functions). There are designated simplex calling frequencies to help hams connect with one another. These are 146.52 MHz (VHF) and 446.00 MHz (UHF), which are intended for FM communication. For hams wishing to use single sideband (SSB) modulation, the calling frequencies are 144.20 (VHF) and 432.10 (UHF). These calling frequencies are part of the "band plan" in the U.S., a voluntary division of bands to avoid interference between incompatible modes. See *www.arrl.org/band-plan* for more information.

B.   HF Bands

If you want to get on the air and connect with other hams, it is important to develop an understanding of the ham radio bands. Look to 20, 40, and 80 meters for most of the high frequency (HF) action, but there are a number of other HF bands to explore, as well.

Hams with a Technician class license are rather limited on the HF bands. Knowing CW (Morse code) opens up portions of the 80-, 40-, and 15-meter bands. Technicians have some phone (voice) privileges on 10 meters and the full range of privileges on 6 meters (the latter being a VHF band).

A ham must decide what band(s) to use at particular times of day and night and with particular operating conditions and equipment. Propagation of the radio signal is a huge consideration. The effectiveness of one's antenna for a particular band is another big determinant. For example, I have an antenna that transmits very well on 40 meters, perhaps somewhat less effectively on 20 meters, and somewhat less well again on 80 meters. Consequently, I tend to favor 40 meters even during daytime hours when 20 meters otherwise has better propagation, unless I am trying to reach Europe. I rarely use 80 meters except for contests or if I happen to notice on a spotting website (such as DXMAPS.com) that an interesting DX station is making contacts.

Propagation evolves throughout the day and night. A key issue is whether the person or area you want to connect with over the air shares your daylight or darkness. After sunrise in North America, Europe remains in daylight, but late afternoon in North America means darkness prevails in Europe. Your best bet for DX contacts is to find times of day when you and your target audience are either both in daylight or both in darkness.

The DXMAPS "QSO/SWL real time QSOs" website (*www.dxmaps.com*) provides an excellent illustration of how propagation changes throughout the day. Check out the two maps below, both from around 8:45 a.m. CST (14:45 UTC). The first one shows SSB and CW QSOs on 40 meters, and most of the contacts are confined within an individual continent (*intra*-continental QSOs). The propagation is not good for DX'ing.

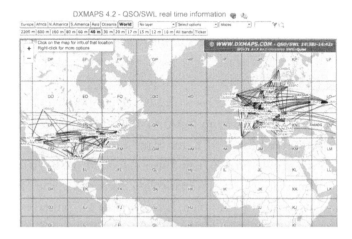

The next one shows SSB and CW QSOs on 20 meters, which has many contacts being made across the Atlantic Ocean (*inter*continental QSOs). If you are looking to make DX contacts during daytime, clearly 20 meters is favored over 40.

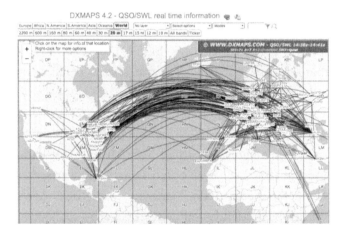

You don't need a computer-generated map to know this. Get on 20 meters on a Saturday morning in the U.S., and chances are you will hear people calling from Europe. Switch to 40 meters and, unless you have a remarkable antenna, I doubt that you will hear European stations (at least that is my personal experience with the equipment I have, and the map above strongly confirms that).

During the evening in the U.S., the opposite is true: 20 meters tends to goes dead, but 40 meters comes alive with DX stations.

80 Meters (3.5 MHz)

The 80-meter band spans from 3.500 to 4.000 MHz, with the portion below 3.600 reserved for RTTY and data (including CW). For single sideband (SSB) phone (voice), use lower sideband (LSB) (which your radio may or may not select automatically). Sometimes, the upper part of the band, used mostly for voice, is referred to as 75 meters.

Eighty meters tends to be more of a short-distance / regional band—good for groups of friends to chat on. Because of high D layer absorption until sunset, 80 meters usually is only good for local communications during the day.

Eighty meters served me well during the 2021 Minnesota QSO Party, where I made 35% of my 229 QSOs on 80 meters. Most of those contacts were to fellow Minnesotans, who I have a harder time connecting with on 40 and 20. The band also worked for contacts to the neighboring states of WI, IA, SD, MI, and IL. (The 80-meter contacts that day were not all local: I was able to reach as far as West Virginia and Maine.)

I have had some, but limited, DX success on 80 meters. For example, I was able to have a QSO with a guy in Serbia at 3:44 UTC. We gave each other signal reports of 5x5, which is not bad given the distance and the fact that I was only running 100 watts. Also, I have had QSOs on 80 meters with hams in Costa Rica, Alaska, and a number of Caribbean islands—all with 100 watts.

Technician class licensees who know CW are permitted to operate CW only from 3.525 to 3.600.

40 Meters (7 MHz)

The 40-meter band (7.000 to 7.300 MHz) is one of the most popular amateur bands. As with 80 meters and lower, use lower sideband (LSB) for phone (voice). In the U.S., Amateur Extras can operate from the bottom of the band to 7.125 MHz using RTTY and data (including CW), with 7.125 to 7.300 open for phone and image use. General class licensees have a somewhat narrower range to work, with phone privileges starting at 7.175 on the bottom end, for example. It behooves Technician class licenses to learn CW, because 7.025 to 7.125 is open to them for this purpose. Technicians do not have any phone privileges on this band.

Forty meters has been my favorite band, although that may be in large part because my ZS6BKW antenna seems to perform best on this band. I get good propagation around the U.S. during the daytime on 40 meters, and good DX propagation at night. For example, from my QTH in Minnesota, throughout the winter I usually hear one or more hams in South Africa just about any night by tuning around the lower end of the single sideband portion of 40 meters (7.128-7.175), and I have made several contacts with hams there.

I've noticed excellent propagation (and people taking advantage of it) on 40 meters to Europe on winter Saturdays around sunset (say 5:30 pm CST), at which time it is 11:30 p.m. or 12:30 a.m. in much of Europe. The hams seem to be up and on the air then, because they don't have to get up early for work the next day.

In the warmer months, 40 (and 80) meters can be plagued by static crashes from distant (and nearby) electrical storms, whereas 20 meters seems to remain quiet.

20 Meters (14 MHz)

Twenty meters (14.000 to 14.350 MHz) is a popular high frequency band for amateur use (although Technician class licensees have no privileges on this band). The lower portion of the band (up to 14.150) is reserved in the U.S. for RTTY and data (including CW), with the portion above that open for phone and image use. General class licensees can operate SSB from 14.225 to 14.350. Twenty meters and higher HF bands use upper sideband (USB) for phone.

A lot of DX (long distance) contacts take place here. A long-time ham told me that 20 meters has always been THE band for DX. (44% of my total QSOs have been made on 20 meters, compared with 40% on 40 meters, 10% on 80 meters, etc.) Twenty meters is desirable from antenna perspective, too: a half-wave dipole only needs to be about 10 meters wide, and beam antennas for this band are not gargantuan (which is an issue with a 40-meter beam).

In my experience, 20 meters primarily is a daytime band. It can be a great band for DX contacts to Europe after sunrise for North American hams and before the sun has set in Europe. During winter in the northern U.S., for example, the sun may not rise until 7:00 or 7:30 a.m. That is when you might hop on 20 meters if you want to reach across the Atlantic Ocean. By 11:00 a.m. CST, the sun may have set in most of Europe, diminishing propagation.

At night during my first several months operating as a ham, I generally would hear nothing on 20 meters. That is until springtime when I started noticing hams in New Zealand and Australia talking to American hams around 10:00 p.m. Central Daylight Time (mid-afternoon local time in New Zealand). (I do not know if there actually was a seasonal change of activity or if it merely was a coincidence that I started noticing nighttime 20-meter activity that time of year—I have not been a ham long enough to figure that out.) After a lot of attempts trying to get through the pileup with 100 watts and no beam antenna, I managed to be picked up by a Kiwi, but it required tremendous persistence. I heard a number of other guys with only 100 watts make contact, but primarily that was due to that particular Kiwi periodically asking the high-power boys to stand down while the 100-watt folks made a call.

A lot of 20-meter DX action seems to occur in the Amateur Extra and Advanced SSB portion of the band (14.150-14.225), which is a good reason to get an Amateur Extra license.

17 Meters (18 MHz)

For almost the first full year I was licensed, I ignored the 17-meter band (18.068 to 18.168 MHz, with 18.110 up for SSB), tuning across it a few times but never attempting to make contacts. In retrospect, it was a big missed opportunity, as it's now one of my favorite bands for DX. For example, over a holiday weekend of sporadic, casual operating, I was able to pick up two new DX entities (Western Sahara and Sardinia) as well as make contacts in Spain, Luxembourg, Belgium, Portugal, Italy, and Mexico—all "barefoot" with 100 watts. My ZS6BKW (dipole) antenna works great on this band (although very poorly on the nearby 15-meter band).

The propagation on 17 is similar to 20 meters—a solid daytime band. No contesting is allowed on 17 meters, which keeps this narrow (100 kHz wide) band quieter. Use USB for phone mode.

15 Meters (21 MHz) and Other HF Bands

The 15-meter band (21.000 to 21.450) is suitable for long-distance (DX) communications. The band is most useful for intercontinental communication during daylight hours, especially in years close to solar maxima. The 15-meter wavelength is harmonically related to that of 40 meters, often making it possible to use an antenna designed for 40 meters. Like with 20 meters, use USB for phone. The phone portion of the band is 21.200 to 21.450 for Amateur Extras. Phone privileges for General class licensees start at 21.275 and go up from there. This is another band allowing Technicians to use CW (21.025-21.200).

I have limited experience on 15 meters, but it has provided a forum for making more contacts during busy contests, providing a helpful score multiplier. For example, during the ARRL International DX Contest, after I had spun the dial back and forth on 20, 40, and 80 and hearing the same call signs over and over, I moved to 15 meters and picked up some new contacts, as well as connecting with folks that I had talked to on other bands (which, in a contest, count again).

There are additional ham radio HF bands to explore, including 160, 30, 12, and 10 meters. Those with a Technician license will be particularly interested in the 10-meter band, because phone (voice) privileges are granted in the range of 28.300 to 28.500 MHz (200-watt power limit). The ARRL 10-Meter Contest is a great time to try out that band.

6 Meters (50 MHz - The Magic Band)

The 6-meter band (50 MHz) is considered very high frequency (VHF)—the lowest portion of that frequency range allocated to hams, not a high frequency (HF) band. However, occasionally it displays propagation characteristic of the HF bands, normally occurring close to sunspot maximum. Because sporadic E propagation sometimes enables long-distance contacts on 6 meters, including intercontinental communications, many amateur operators refer to 6 meters as the "magic band." May through August is best in the northern hemisphere, whereas November through February is prime time in the southern hemisphere.

In the U.S., hams can operate 50.0-50.1 (CW only) and 50.1-54.0 MHz using RTTY, data, phone and image. 50.125 MHz is the traditional 6-meter single-sideband (SSB) calling frequency. For FM communications, 52.525 simplex is the place to listen. Since it's VHF, Technician class licensees have full operating privileges on this band.

There are VHF contests at various times during the year, such as ARRL's contests in January, June, and September, and the CQ VHF contest in July. These are great opportunities to make contacts on 6 meters.

There is a huge range of bands and operating modes to explore, depending on your operating privileges, your equipment, and your interests. Check 'em out!

9

MODES OF OPERATION

Amateur radio operators can use a variety of communications modes over radio, such as voice, image, and data. A "mode" can be either a modulation mode like FM or AM, or it can be an operating mode such as voice, Morse code ("CW"), or data packets.

A.   Voice Modes ("Phone")

Voice is the easiest way to start communicating on the air. Most every rig comes equipped with a microphone. All you have to do is find an appropriate frequency, tune the antenna if necessary, use the Push-To-Talk (PTT) button, and speak.

On high frequency (HF), most hams use single sideband (SSB), although some use other (older) forms of modulation, such as AM. Communications on VHF and UHF generally use FM (frequency modulation).

In addition to these types of analog modulation, there are digital voice modes such as D-STAR, DMR, and System Fusion. A ham using an analog (FM) VHF/UHF handheld radio tuned to a repeater frequency may hear the squelch break from time to time only to receive static. This may be the result of a digitally-encoded transmissions, as some repeaters process both analog and digital signals. This can be annoying if an analog-only HT is set to scan a range of repeater frequencies and the scan keeps stopping on digital signals. If this happens to you, you might need to remove one or more digital-capable repeaters from the scan sequence.

(See other parts of this book for terminology and etiquette to use on the air, how to call CQ, and information about bands.)

B.   Morse Code ("CW")

International Morse Code

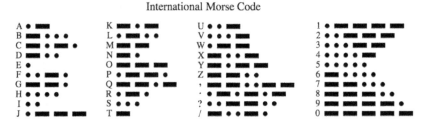

It is important to note that none of the ham radio tests in the U.S. have a Morse code component anymore. Nevertheless, it remains an important operating mode, particularly since it is a very efficient mode from a transmitting-power perspective, with its very narrow bandwidth. Thus, hams using a mere five or ten watts can have tremendous reach, whereas someone using voice over SSB may need 100 watts or more to have comparable reach. Furthermore, it is the only mode that Technician Class licensees are allowed to use on the high frequency bands (for long distance communications), other than a sliver of the 10-meter band where Techs can use voice communications. Techs are granted access for CW to 3.525 to 3.600 on the 80-meter band, 7.025 to 7.125 on the 40-meter band, and 21.025 to 21.200 on the 15-meter band. (Good areas for slow CW include 7.05-7.055, 7.1 to 7.122.) Thus, a Technician who takes the time to learn code will have a powerful way to connect with other hams, including DX stations.

When I was first exploring the idea of learning Morse code for ham radio (before I started studying for a license), I got confused because I kept seeing references to "CW" while researching it online. Some online resources on the subject don't even call it Morse code. "CW" refers to the "continuous wave" transmission used for Morse code, and it has become the stand-in for identifying use of the code on radio.

In the earliest days of radiotelegraphy, Morse code sent by spark-gap transmitter was the first and only wireless communication mode, sending "damped waves" that were very broad and inefficient. Spark-gap transmitters (like the one used on the Titanic to send the distress signal) soon were replaced by continuous wave (CW) transmission, using vacuum tube oscillators capable of generating a pure tone. Communication was accomplished by the operator, using Morse code and a "straight key" to interrupt the continuous transmission into "dits" and "dahs." Radio communication by Morse code was the only way to communicate for the first decade or more of Amateur Radio.

As I mentioned early in this book, I started teaching myself CW in the early 90s using a practice tone generator that I had constructed myself to generate audible tones. With that, I could practice *sending* CW, but not the more important and more difficult task of receiving/decoding it. When you are sending CW, you can never send it faster than your brain can figure out the code. Thus, as the sender, you always are in control. However, when you are receiving it, you have no control over the sender's speed, other than to ask them to slow down after you've already failed to copy what the operator sent.

In 2019 when I took another crack at learning CW, the World Wide Web was available—a resource that did not exist in 1991. With the Internet, suddenly I had access to an abundance of audio recordings of CW at different speeds that I could practice copying, and there are websites that provide a variety of training exercises. One of my favorite sites is "Learn CW Online" (*www.lcwo.net*), which requires registration but is free to use. LCWO uses a combination of Farnsworth timing and the Koch method of study.

Another site that combines those training methods is "Koch method to learn Morse" (*stendec.io/morse/koch.html*). This site is excellent, but it has fewer training options than LCWO. On the plus side, unlike LCWO, this site does not have a pause button, and you can't rewind a lesson. I used it to break my bad habit of stopping the stream of characters and going back to fix missed bits of code, because obviously one can't do that in the real world of radio transmissions. I had to learn to force myself to just keep going and catch the characters that I can and just accept that I will miss some or get some wrong. At least for me, it takes mental discipline to not get stuck on a character that I know but just can't decode in the heat of the moment.

Using Farnsworth timing, the individual characters are sent at high speed with extra spacing inserted between the characters and words to slow the overall transmission down to an understandable pace. The advantage of this is that the student gets accustomed to recognizing characters by sound rather than by counting out the dits and dahs in each character (which is not possible at higher speeds). Thus, it becomes easier to increase overall speed later on.

The Koch method exposes the student to full-speed Morse code from day one, but the first lesson starts with just two characters played at full speed. The student must "copy" them (i.e. write or type them). Once 90% of the characters are correctly "copied," the student can go move to the next lesson, in which just one more character is added. (Some study by adding batches of 4 or 5 letters at a time.)

LCWO employs Farnsworth spacing but also offers the Koch technique of incrementing the number of characters to copy by one, thus incrementally progressing through the alphabet.

ARRL offers code practice MP3 files at eleven different speeds ranging from 5 WPM to 40 WPM, which can be found online at *www.arrl.org/code-practice-files*. ARRL also provides code practice sessions over the air on 20 and 40 meters and on EchoLink at scheduled times. After spending time using the exercises on a site like LCWO (which are useful exercises but not entirely the same as real-world copying), it is helpful to periodically 'truth" one's progress by doing continuous copying from an audio file or from a radio.

Other sites offering CW training and information include:

Learning Morse Code - *www.arrl.org/learning-morse-code*
"AA9PW - Amateur Radio Practice Site" *aa9pw.com/morsecode/*

In addition to using a computer and an Internet connection to learn CW, smartphone apps exist for this purpose, including **Morse-It**, **Ham Morse**, and **Morse Toad**. I have used Morse-It for practicing sending CW and for decoding CW (by setting it next to the radio). It has a Koch trainer feature in the paid version of the app ($4.99 as of this writing) that I have not tried. I would think it would be much more difficult to practice copying code without the benefit of a physical keyboard, though. Morse Toad uses gamification to make learning CW fun. It and Ham Morse also cost $4.99. Apps like these might be good for some, but I'm rarely very far away from my laptop, and I would rather use a free website than a paid app on a device with no keyboard.

Other practice methods I use include translating road signs and billboards into Morse when I am in a car, and sounding out the alphabet in my head when I go to bed at night.

Learning CW is more than knowing Morse code for the alphabet, numbers and punctuation. In addition to that, there is a body of abbreviations and procedure signs ("prosigns") to learn if you want to use the code in the real world. If you had to spell everything out, CW would be much more tedious than it is. Here is a basic list to learn:

Abbrev.	Meaning		Abbrev.	Meaning
AGN	Again		OM	Old Man (term of endearment)
ANT	Antenna		OP	Operator (in place of "name")
BCNU	Be seeing you		PSE	Please
BK	Break-to pause transmission		PWR	Power
C	Yes; correct		QRS	Send Slower
CL	Closing (I am closing my station)		R	Roger
CQ	Calling any station ("seek you")		RIG	Transceiver
CUAGN	See you again		RPT	Report / Repeat please
DE	This is from		RST	Signal report (Readability -Signal Strength -Tone)
DX	Distance (e.g., foreign countries)		SK	Out, end of contact
ES	And		SRI	Sorry
FB	Fine Business (i.e. good)		SVP	Please (French: "S'il vous plaît")
FER	For, for your		TU	Thank you
FWD	Forward		WX	Weather / Weather report follows
GM, GA, GE	Good morning, Good afternoon, Good evening		XYL	Wife (ex young lady)
HH	Correction (8 dits)		YL	Young lady
HR	Here		Z	Zulu time (UTC)
HW?	How are you receiving me?		73	Best regards
N	No; nine		?	Say again

Once you learn CW and the most common abbreviations and prosigns, to use it you also will need to understand how to make contact with other hams (have a QSO). Here is a typical format for a basic CW QSO:

CQ call / answer
RST (signal report) / Name / QTH (location)
Rig (radio) / Power (in watts) / Antenna / WX (weather)
Thanks / QSL info (e.g., direct, buro, or LoTW) / 73

That looks easy enough, right? Well, when you see it all spelled out, it's a bit more intimidating. Fortunately, however, if you learn this basic flow of information, you will have a better idea of what to expect when copying code from another operator.

Following is a sample script. Before calling CQ, it is highly recommended that you listen first, and even if things sound quiet, send "QRL?" or "QRL DE [your call sign]?" to make sure the frequency is clear.

Code	Interpretation
CQ CQ CQ DE N9CD N9CD	"Seek you" (Hey anybody out there?) from N9CD. Over to whoever may answer.
N9CD DE W0VTT W0VTT	N9CD from W0VTT. (Call sign repeated to make sure receiver gets it.)
W0VTT DE N9CD = GM TNX FER CALL BT UR RST 599 599 = NAME HR TIM TIM QTH LACRESCENT MN LACRESCENT MN = HW? N9CD	W0VTT from N9CD, Good morning. Thanks for call. Your signal report is 599. Name here is Tim. Location La Crescent MN. How are you receiving me?
GM TIM TNX FB REPT UR RST 579 579 = NAME IS MIKE MIKE QTH ST CHARLES, MN = HW? DE W0VTT	Good morning, Tim. Thanks for the good ("Fine Business") report. Your signal report is 579. My name is Mike, my location is St. Charles, MN. How are you receiving me?
DE N9CD = R R TNX FER RPT = RIG IS IC706 PWR 100W ANT IS ZS6BKW = WX IS CLOUDY ES COLD 29F = HW?	From N9CD, Roger. Thanks for report. My rig is an IC-706, power 100 watts, antenna is a ZS6BKW [a type of dipole]. Weather is cloudy and cold 29 degrees

	Fahrenheit. How do you copy?
R R RIG IS K3S 100W ES ANT IS DIPOLE = WX IS WET AND COLD HW?	Roger. My rig is an Elecraft K3S, 100 watts, and antenna is a dipole. Weather is wet and cold.
W0VTT DE N9CD = FB TNX FER QSO = PSE QSL VIA BURO = 73 ES CU AGN AR N9CD SK	W0VTT from N9CD. Fine business. Thanks for QSO [contact]. Please QSL via the bureau. Best wishes and see you again. End of message, over to you.
N9CD DE W0VTT = FB TIM TU FER NICE QSO 73 ES BCNU AR N9CD DE W0VTT SK CL	N9CD from W0VTT, Fine business Tim. Thank you for the nice QSO. Best wishes and be seeing you. End of message, end of contact, I'm closing my station.
TU 73	Thank you. Best wishes.
TU E E	Thank you. (Two dits is a customary sign-off. The other party may respond with a single dit.)

The double hypen is used as a spacer between thoughts or subjects. The code for = is sent in CW as dah dit dit dit dah, which happens to be the same as the CW letters B and T sent together. (Usually it is written out with a line over the top of the letters.) Some folks will use a period instead.

If you make a mistake, sending 8 dits in a row (HH) is the way to communicate that, then correct your mistake.

When answering a CQ, you should "zero beat" the caller's frequency (set your transmit frequency as close to theirs as possible). On most modern radios, if the pitch your side tone matches the pitch of the other guy's signal in your receiver, you are on the same frequency. Also, when calling CQ, do so with your narrow CW filter turned off, or you might not hear answering hams.

I have made progress with my CW skills using the stendec.io and LCWO sites and ARRL on-air practice sessions and online audio files, but I have a lot more practice to do. I look forward to trying to make

QSOs with CW. I feel that learning CW will greatly enhance my overall operating skills and open up new opportunities for QRP operation.

C.   Digital Modes

Digital modes involve connecting a computer to an amateur radio, except in the case of Morse code (CW) which, traditionally, uses the human brain as the computer to encode and decode signals (although this can be done with digital computers now, too).

There are some huge benefits to using digital modes, including the ability to connect with other hams using extremely weak signals, or sending pictures or videos. Digital modes have become very popular, and use of digital modes is growing rapidly.

To use digital modes with older ham radios, one merely needs a relatively inexpensive interface device to connect the radio with a computer, such as SignaLink or RIGblaster, and software to handle the Digital Signal Processing (DSP). The setup may employ the computer's soundcard or the interface device might have its own soundcard.

RTTY

RTTY (pronounced "ritty") a/k/a "radioteletype" is referred to as the original data mode and has been around since the first half of the 1900s. Unlike most digital modes, RTTY is transmitted on lower sideband (LSB). Unfortunately, RTTY doesn't perform particularly well at very weak signal levels.

PSK

Phase Shift Keying (with various flavors) has become one of the most popular of the newer digital modes, probably because it excels at pulling data from weak signals. PSK31 has a bandwidth of only 60 Hz. Thus, many signals can fit into the same bandwidth that would be occupied by a single sideband (SSB) signal. PSK31 is a keyboard-to-keyboard mode, similar to RTTY, and works well for having a nice QSO with another ham—something that is not possible on all digital modes.

FT8

An August 2017 article from ARRL (already ancient news, I realize) bears a headline calling the FT8 Mode "the Latest Bright Shiny Object in Amateur Radio Digital World." It allows hams to complete QSOs when signals are too weak for traditional modes. It provides perhaps as much as a 20 dB advantage over SSB and 10 dB over CW. To maximize the chances of success, each FT8 packet holds 13 characters and takes 13 seconds to send. I have heard it compared to Twitter in its brevity. It is not useful to anyone who wants to ragchew, but it does allow for making a lot of contacts.

D.   Images and Video (Amateur TV)

Amateur radio operators can send images through the airwaves: slow-scan TV (SSTV) on high frequency (HF) bands (still pictures), and fast scan television (FSTV) on VHF/UHF bands (video).

Slow Scan TV has been popular for many years. The received still pictures are built up line-by-line over the course of about a minute—not like what we are used to these days with sending images over broadband Internet. SSTV usually only takes up to a maximum of 3 kHz of bandwidth, compared to analog broadcast television, which requires at least 6 MHz wide channels.

Unlike the grainy still images of SSTV, fast scan television (a/k/a amateur television or ATV) involves the transmission of broadcast quality video and audio. In North America, amateur radio bands that are wide enough to fit a television signal are higher in frequency than VHF broadcast TV. The lowest frequency ham band suitable for ATV transmission is 70 centimeters--between broadcast channels 13 and 14. Other bands are also used for amateur television, mostly in the UHF region on frequencies higher than UHF broadcast TV, such as 33 and 23 centimeters. I have heard that ATV is useful for many public service and emergency comms events, but I have yet to try amateur television or see it in action.

E.   QRP vs. High Power

Marketing photo for the Lab599 Discovery TX-500 QRP rig.

When I first started preparing to take the amateur radio tests, I was very interested in QRP (low power) operation. I wanted to be able to take a rig backpacking. After all, the primary motivation that pushed me to dive into ham radio (other than being interested in learning Morse Code) was a trip I had planned to the Boundary Waters Canoe Area Wilderness, where I was concerned that connections to the rest of the world might be difficult or impossible. I saw images online of people with QRP rigs off in the mountains and wilderness, and that looked very appealing. It just so happened that as I was preparing to get licensed, a couple of new and very impressive-looking QRP rigs were slated to be released within months: the Icom 705 and the Lab599 Discovery TX-500.

On the flipside of these advanced rigs with waterfall displays and built-in digital modes, there's also the appeal of the quintessential homebuilt CW rig housed in an Altoids tin—ham radio stripped to its bare essence—raw and powerful in its utter simplicity.

As appealing as remote QRP operation initially seemed to me, the more I read and heard from other hams, the less interested in QRP I became as a starting point for getting into ham radio. Experienced guys told me I would get frustrated, especially trying to operate single-sideband (SSB) voice on very low power. Sometimes it is hard enough making contacts with a full-power, 100-watt rig. Why limit oneself to five or ten watts when starting out? (QRP has come to mean five watts or less output for CW, or ten or fewer watts PEP output for SSB.)

It is possible to have a single rig that can be used out in the field, mobile in the car, and at home as a base station. One of my Elmers tells me his Elecraft KX3 transceiver coupled with a KXPA100 100 watt amplifier work great for a base rig, portable and even mobile. I suppose a ham could do something similar with the new IC-705. However, those are not inexpensive starter options.

Unless your only interest is operating out in the field or only doing CW, you probably will want to start with a 100-watt base station and build experience and confidence with that.

Speaking of 100-watt transceivers, I would be remiss to not mention the Icom IC-7300. I understand that it has changed the face of ham radio in recent years. It's an HF rig that packs many features into a small box, often for under $1,000 (and a lot are available used for even less). The rig covers all of the HF bands plus 6 meters, it has a built-in automatic antenna tuner, and it sports a digital display with a high-resolution waterfall function.

My first HF rig is an Icom IC-706MKIIG—an older, discontinued model. It is referred to as a "DC to daylight" rig, because it covers such a huge range of the electromagnetic spectrum: 160 meters on the low end all the way up to the 70 cm band (UHF). It provides base station performance (100 watts) and features in a mobile rig-sized package. When I eventually get a new base station, I hope to move the IC-706 into my truck for mobile operation.

As with all aspects of amateur radio, I do encourage you to learn more and make independent decisions about whether a particular facet of the hobby is right for you. To learn more about QRP, check out *www.arrl.org/why-qrp* and the many other online resources and YouTube videos on the subject.

I do hope to eventually acquire a QRP rig and use it out in the field, especially when I get good enough with CW to be able to engage in QSOs with a Morse code paddle rather than a microphone. But for now, I'm enthralled with making SSB contacts around the country and around the world and working toward some awards.

10

CONCLUSION

Amateur radio is rewarding in a multitude of ways. It is all about the journey: building contacts and experiences while making friends and sharpening skills. Similar to other gear-dependent, but skill-based and licensed activities like flying airplanes and scuba diving, there is a nearly infinite variety of experiences to enjoy and skills and equipment to build or add. Whether you are into the gear itself, exploring different modes of operation, reaching farther and farther around the globe, making contacts, earning awards, connecting with local club members, or all of the above, there always is a new adventure in ham radio to be savored. I never expected to enjoy it nearly as much as I do.

At some point in pursuing ham radio, you will experience a hiccup or two (or ten). I certainly did.

For example, when I bought my new handheld VHF/UHF radio, the programming cable would not work (a common problem). Thus, I could not program any frequencies into it except through a tedious manual process. I was discouraged. Fortunately, the local ham radio club president helped me out by experimenting with me and then loaning his cable to me.

Later, about half a year after I started on HF, my rig suddenly stopped receiving any signals. I did not know whether my used rig malfunctioned or whether it was an antenna issue, and I did not own any diagnostic equipment. I pulled the dipole antenna out of the trees (not an easy process in my case), but I could not find any breaks in the antenna. Eventually, after being off the air for quite a long time, I discovered that one of the PL-259 connectors on the balun connecting the feed line to the antenna had a broken connection. After a bit of soldering, I was back in business and immediately made several DX contacts!

Also, even though there are huge numbers of wonderful and helpful people in ham radio, occasionally you will run into jerks on the air. Don't let a few grumpy old men discourage you.

An acquaintance of mine, Ralph Heath, published a book entitled, *Celebrating Failure: The Power of Taking Risks, Making Mistakes, and Thinking Big*. Heath, like anyone who has studied successful people, knows that they all fall down at some point. The measure of their success is not how seldom they fall; rather, how well they pick themselves up and keep going, striving even harder to succeed.

Whether you find yourself struggling with fickle hardware, complicated exam questions, or learning CW, take comfort in knowing that every other ham has struggled in some ways, too. Fortunately, I will almost guarantee there is an Elmer willing to help you overcome that hurdle. Or there's a YouTube video, a book, or an article explaining the subject.

Relish the journey that awaits you in amateur radio. *Bon voyage* and 73!

APPENDIX A

ONLINE RESOURCES TO HELP YOU ON YOUR HAM RADIO JOURNEY

Here are links to some websites to help you on your journey exploring the world through ham radio:

Amateur Radio License Map - *haminfo.tetranz.com/map*

American Radio Relay League - *ARRL.org*

ATV 101 – An Introduction to Fast Scan Amateur Television FSTV - *www.hamuniverse.com/atvfastscantv.html*

DX Engineering (equipment) - *www.dxengineering.com*

DXMAPS real time QSOs - *www.dxmaps.com*

eQSL – electronic QSL card exchange – *www.eQSL.cc*

Gigaparts (equipment) – *www.gigaparts.com*

Ham Radio Classified Ads – *www.swap.qth.com*

Ham Radio Crash Course – *www.hrcc.stream* (website and YouTube channel)

Ham Radio Outlet (equipment) - *www.hamradio.com*

HamRadioQRP - videos about learning CW online, etc. - *hamradioqrp.com*

Learn CW (Morse code) Online – *www.lcwo.net*

License Exam Practice for Ham Radio - *www.arrl.org/exam-practice*

License prep and education service - *HamRadioSchool.com*

Mac Ham Radio Links - *www.machamradio.com*

Mac for Ham Radio - Why Go Mac - *www.qsl.net/ah6rh/am-radio/mac/*

Practice Amateur Radio Exams - *www.qrz.com/hamtest/*

Parks on the Air – realtime call signs & frequencies of hams operating from parks across the country – *pota.us*

QRZ.com - Looking up call signs, logging contacts, QSLing, obtaining awards

QSL Card Creator - *www.radioqth.net/qslcards*

Radio Reference - claiming to be the world's largest radio communications data provider – *www.RadioReference.com*

Remote rigs accessible through *www.RemoteHams.com*

Repeaterbook - claiming to be Amateur Radio's most comprehensive, worldwide, FREE repeater directory - *www.repeaterbook.com*

Satellites for Amateur Radio – *www.n2yo.com/satellites/?c=18* *www.amsat.org*

APPENDIX B

GLOSSARY

73	Best wishes (used on the air and in correspondence)
AM	Amplitude Modulation
APRS	Automatic Packet Reporting System (uses GPS)
ATV	Amateur Television, a/k/a FSTV (Fast Scan Television)
Balun	A device used for coupling a balanced load with an unbalanced load, such as connecting coax feed line to a vertical length of ladder line on certain types of dipole antennas
Barefoot	Running a transmitter without an amplifier
Bug	Mechanical keying device for sending Morse code semi-automatically
Bureau or Buro	Organization that provides a collecting and distributing point for QSL cards
CQ	"Seek you" - a way to initiate contact with anyone who may be listening
CTCSS	Continuous Tone-Coded Squelch System - transmitting a tone along with radio signal to instruct a receiver to open up its squelch so that the signal is received, a method commonly used with repeaters
CW	"Continuous Wave" - a mode for sending messages by Morse Code
DE	"from" – used in CW
Dipole	The simplest and most widely used class of antenna, commonly consisting of two identical

	conductive elements such as metal wires or rods connected to a radio transceiver with a feed line.
Dummy load	Power-dissipating device substituted in place of an antenna on a transmitter, used for testing
DXCC	"DX Century Club" - ARRL-sponsored award for making contact with 100 different countries by ham radio
DXing	"Distance" - the pursuit of distant stations, generally hams in foreign countries
DXpedition	Trip made by ham operator(s) to a "DX" country for the purpose of providing other hams an opportunity to make contacts into that country
Elmer	A mentor for a prospective or new ham operator
EME	Earth-Moon-Earth communications, a/k/a moon bounce
FB	"Fine business" (good)
FM	Frequency Modulation
Green stamp	U.S. dollar bill sent with a QSL card to pay postage costs for a return card
HF	High Frequency - the realm of most long distance amateur radio communications between 3 and 30 megahertz (MHz)
HT	Handy-Talkie - a handheld transceiver, usually for VHF and UHF
J-Pole	A type of antenna typically used for VHF/UHF
LSB	Lower Sideband - a voice mode typically used on the 40-, 80-, and 160-meter HF bands
Paddles	Device used for sending CW, connected to a keyer
PEP	Peak Envelope Power
PSK31	A digital mode using phase shift keying
QRM	Interference from another station
QRN	Interference from static
QRP	Low power
QRZ	(Spoken as "Q-R-Zed") Who is calling me?
QSB	Your signals are fading
Q signals	Shorthand abbreviations for CW, some of which get used with voice communications, too
QSL	Two meanings: (1) acknowledge receipt of a

	transmission; (2) a card providing written acknowledgment that a contact has been made
QSO	An over-the-air contact or conversation
QTH	A ham operator's location
Ragchewing	Extended, informal conversation, in contrast with making rapid radio contacts with large numbers of ham operators such as in a contest
Repeater	Device that retransmits a signal and can greatly extend the range of a handheld VHF/UHF transceiver
Rig	Transceiver
RST	"Readability, Strength, Tone" - system for providing signal reports, used for CW. For SSB, only the R and S are used (e.g. 5-9)
RTTY	Radio teletype
SSB	Single-Sideband - a type of modulation, generally used for voice
SSTV	Slow-Scan Television – a method for transmitting still images
SWR	Standing Wave Ratio - a measure of impedance matching, used to determine suitability of an antenna
VHF	Very High Frequency (30-300 MHz) - frequency range generally used for shorter, line-of-sight communications an for amateur satellite contacts
UHF	Ultra High Frequency (300 MHz to 3 GHz) - frequency range for line-of-sight communications and for amateur satellite contacts
USB	Upper Sideband - a voice mode typically used on HF bands from 20 meters up
VEC	Volunteer Exam Coordinator – one who administers amateur radio examinations
WAS	"Worked All States" award
WX	Weather
XYL	One's wife (ex-YL or ex Young Lady)
Yagi-Uda or Yagi	Type of directional or beam antenna that provides signal gain (named after its Japanese inventors)
YL	Young Lady – a female

APPENDIX C

Study Guide and Test Questions for Technician Class License

In the coming pages, you will find every single question that could appear on the Technician Class license exam along with the correct answer only.

Many exam prep materials present four possible multiple-choice answers. Nearly seventy-five percent of the answers presented are the incorrect responses. (Some of the answers are "all of the above," in which case the four answers presented are all correct, making the number of incorrect answers presented in the pool somewhat less than 75%.) Seeing all of those wrong answers will fill your brain with extra and incorrect information. Save time and brain cells by studying only the correct answers. That way, on test day, the wrong answers will appear unfamiliar, and you will be less likely to choose them.

The entry-level amateur radio license is Technician class. If you earn a Technician license, you will be in good company: as of early 2019, Technician licensees comprised more than half of the US Amateur Radio population, numbering around 384,500.

The Technician test is relatively easy for most people. There are only 35 questions to answer (out of 423 possibilities). According to the ARRL as of 2018, the pass rate is 81%. The Entry-Level License Committee, which is proposing to have the FCC expand the privileges of this license, determined that the current Technician class question pool already covers far more material than necessary for an entry-level exam to validate expanded privileges.

From a studying perspective, there are a few primary types of questions on the ham radio tests, including (1) regulatory questions that are entirely arbitrary that you simply will have to memorize; (2) self-evident or nearly self-evident questions that most people can figure out just by looking at the multiple choice answers and eliminating the ones that seem clearly wrong; and (3) questions of science/electronics, some of which have to be memorized and others that are subject to calculation based upon simple formulas.

An example of a regulatory question with an arbitrary answer is, "What is the limitation for emissions on the frequencies between 219 and 220 MHz?" The answer (fixed digital message forwarding systems only) is not something of common public knowledge, and it is not something you can compute. Just read the question and answer and hope it sticks.

Here is an example of the fairly self-evident type of question:

Q. Which agency regulates and enforces the rules for the
   Amateur Radio Service in the United States?
A. The FCC

If you were inclined to answer that one with the "Bureau of Alcohol, Tobacco, Firearms and Explosives," maybe the amateur radio test isn't for you after all. Of course, that is not one of the multiple-choice options anyway, although "Homeland Security" is listed as an [incorrect] option.

Here's another fairly self-evident one:

Q. How many operator/primary station license grants
   may be held by any one person?

A. One

One person—one license. Pretty straightforward, right? Even if you did not know the answer before, it should not be hard to remember now.

Then there are questions that require the test-taker to perform a calculation:

How much power is being used in a circuit when the applied voltage is 13.8 volts DC and the current is 10 amperes?

A. 138 watts
B. 0.7 watts
C. 23.8 watts
D. 3.8 watts

For this, you will need to know one of the few formulas necessary for the test, specifically that voltage multiplied by amperage equals power (measured in watts). In the question above, multiply the 13.8 volts by the 10 amps to get 138 watts. The test writers intentionally made the math easy: multiplying any number by ten simply requires moving the decimal point one digit to the right, transforming 13.8 to 138.0. Usually, the wrong answers on the math problems are way off, which means you can calculate an estimated answer without a calculator and still select the correct response (although you are allowed to use a calculator).

Don't worry—you can do this! Just review the test questions and answers, learn and practice a few formulas, and take some practice tests. Based upon the results of the practice tests, you will know when you are ready to tackle the real thing.

Taking Practice Tests

One you have studied the test material, you should take practice tests to determine how well you have retained the material and to decide when you are ready to attend a test session. Fortunately, there are websites with free access to practice tests using the actual questions on the exam, which are randomly sorted. The online tests will give you an immediate test score, and at least some will track how many questions in the test pool you have encountered as you take the test multiple times.

I used the practice tests on the QRZ.com website: *www.qrz.com/hamtest*. You have to register an account on the site, but it is free. Ham Study (*hamstudy.org*) is another site for taking practice tests.

Taking the Actual Test

When you are ready to take the actual test, you can do so in person or online. ARRL has a "Find an Amateur Radio License Exam in Your Area" search feature at *www.arrl.org/find-an-amateur-radio-license-exam-session*.

On exam day, you will need to bring one legal photo ID or, if no photo ID is available, two forms of identification. (Requirements for minors are somewhat different.) In advance of the test, you will need to obtain a Federal Registration Number (FRN) (*see www.fcc.gov/wireless/support/universal-licensing-system-uls-resources/getting-fcc-registration-number-frn*). New license applicants should create an FCC user account and register in the FCC Commission Registration System (CORES). If you already have an amateur radio license, you will need to present a copy. Bring a couple of number two pencils with functional erasers and a pen. You are allowed an electronic calculator, as long as it does not have any formulas stored in memory. Don't forget to bring the examination fee (which you may need to pay in cash for in-person tests), and know your e-mail address.

You may find out within minutes as to whether you passed the Technician test. If you did, go ahead and take the general test, too. (Hopefully you will have studied for that, as well.)

I you have passed one or more tests, you will be able to get on the air and transmit as soon as your license shows up in the FCC database: *https://wireless2.fcc.gov/UlsApp/UlsSearch/searchAmateur.jsp*. For me, it took ten days (a *looong* ten days). Fortunately, that gave me time to do some radio shopping.

The test questions and answers begin on the next page. Good luck!

FCC Exam Question Pool for Technician Class License

Effective 7/1/2018 - 6/30/2022

SUB-ELEMENT T1 – FCC Rules, descriptions, and definitions for the Amateur Radio Service, operator and station license responsibilities - [6 Exam Questions - 6 Groups]

T1A - Amateur Radio Service: purpose and permissible use of the Amateur Radio Service, operator/primary station license grant; Meanings of basic terms used in FCC rules; Interference; RACES rules; Phonetics; Frequency Coordinator

Q. Which of the following is a purpose of the Amateur Radio Service as stated in the FCC rules and regulations?

A. Advancing skills in the technical and communication phases of the radio art

Q. Which agency regulates and enforces the rules for the Amateur Radio Service in the United States?

A. The FCC

Q. What are the FCC rules regarding the use of a phonetic alphabet for station identification in the Amateur Radio Service?

A. It is encouraged

Q. How many operator/primary station license grants may be held by any one person?

A. One

Q. What is proof of possession of an FCC-issued operator/primary license grant?

A. The control operator's operator/primary station license must appear in the FCC ULS consolidated licensee database

Q. What is the FCC Part 97 definition of a beacon?

A. An amateur station transmitting communications for the purposes of observing propagation or related experimental activities

Q. What is the FCC Part 97 definition of a space station?

A. An amateur station located more than 50 km above the Earth's surface

Q. Which of the following entities recommends transmit/receive channels and other parameters for auxiliary and repeater stations?

A. Volunteer Frequency Coordinator recognized by local amateurs

Q. Who selects a Frequency Coordinator?

A. Amateur operators in a local or regional area whose stations are eligible to be repeater or auxiliary stations

Which of the following describes the Radio Amateur Civil Emergency Service (RACES)?

A. A radio service using amateur frequencies for emergency management or civil defense communications
B. A radio service using amateur stations for emergency management or civil defense communications
C. An emergency service using amateur operators certified by a civil defense organization as being enrolled in that organization
D. All of these choices are correct

Q. When is willful interference to other amateur radio stations permitted?

A. At no time.

T1B - Authorized frequencies: frequency allocations; ITU; emission modes; restricted sub-bands; spectrum sharing; transmissions near band edges; contacting the International Space Station; power output

Q. What is the International Telecommunications Union (ITU)?

A. A United Nations agency for information and communication technology issues

Q. Which amateur radio stations may make contact with an amateur radio station on the International Space Station (ISS) using 2 meter and 70 cm band frequencies?

A. Any amateur holding a Technician or higher-class license

Q. Which frequency is within the 6 meter amateur band?

A. 52.525 MHz

For the question above, and any time you want to convert a frequency to a band (designed by wavelength) or vice versa, divide 300 by the number. In this case, 300 divided by 52 MHz is 5.7 (which you can round up to 6, as in the 6-meter band). If the frequency was 7.250, divide 300 by that, and you will get about 41 (round to get the answer: 40-meter band). Or if you know the band (e.g., 80 meters), divide 300 by that, which equals 3.75 as in 3.75 MHz, which is part of the 80-meter band (3.600 to 4.000 MHz). The following question uses the same formula:

Q. Which amateur band are you using when your station is transmitting on 146.52 MHz?

A. 2 meter band

Q. What is the limitation for emissions on the frequencies between

219 and 220 MHz?

A. Fixed digital message forwarding systems only

Q. On which HF bands does a Technician class operator have phone privileges?

A. 10 meter band only

Q. Which of the following VHF/UHF frequency ranges are limited to CW only?

A. 50.0 MHz to 50.1 MHz and 144.0 MHz to 144.1 MHz

Q. Which of the following is a result of the fact that the Amateur Radio Service is secondary in all or portions of some amateur bands (such as portions of the 70 cm band)?

A. U.S. amateurs may find non-amateur stations in those portions, and must avoid interfering with them

Q. Why should you not set your transmit frequency to be exactly at the edge of an amateur band or sub-band?

A. To allow for calibration error in the transmitter frequency display
B. So that modulation sidebands do not extend beyond the band edge
C. To allow for transmitter frequency drift
D. All of these choices are correct

Q. Which of the following HF bands have frequencies available to the Technician class operator for RTTY and data transmissions?

A. 10 meter band only

Q. What is the maximum peak envelope power output for Technician class operators using their assigned portions of the HF bands?

A. 200 watts

Q. Except for some specific restrictions, what is the maximum peak envelope power output for Technician class operators using frequencies above 30 MHz?

A. 1500 watts

T1C - Operator licensing: operator classes; sequential and vanity call sign systems; international communications; reciprocal operation; places where the Amateur Radio Service is regulated by the FCC; name and address on FCC license database; license term; renewal; grace period

Q. For which license classes are new licenses currently available from the FCC?

A. Technician, General, Amateur Extra

Q. Who may select a desired call sign under the vanity call sign rules?

A. Any licensed amateur

Q. What types of international communications is an FCC-licensed amateur radio station permitted to make?

A. Communications incidental to the purposes of the Amateur Radio Service and remarks of a personal character

Q. When are you allowed to operate your amateur station in a foreign country?

A. When the foreign country authorizes it

Q. Which of the following is a valid call sign for a Technician class amateur radio station?

A. K1XXX

Q. From which of the following locations may an FCC-licensed amateur station transmit?

A. From any vessel or craft located in international waters and documented or registered in the United States

Q. What may result when correspondence from the FCC is returned as undeliverable because the grantee failed to provide and maintain a correct mailing address with the FCC?

A. Revocation of the station license or suspension of the operator license

Q. What is the normal term for an FCC-issued primary station/operator amateur radio license grant?

A. Ten years

Q. What is the grace period following the expiration of an amateur license within which the license may be renewed?

A. Two years

Q. How soon after passing the examination for your first amateur radio license may you operate a transmitter on an Amateur Radio Service frequency?

A. As soon as your operator/station license grant appears in the FCC's license database

Q. If your license has expired and is still within the allowable grace period, may you continue to operate a transmitter on Amateur Radio Service frequencies?

A. No, transmitting is not allowed until the FCC license database shows that the license has been renewed

T1D - Authorized and prohibited transmission: communications with other countries; music; exchange of information with other services; indecent language; compensation for use of station; retransmission of other amateur signals; codes and ciphers; sale of equipment; unidentified transmissions; one-way transmission

Q. With which countries are FCC-licensed amateur radio stations prohibited from exchanging communications?

A. Any country whose administration has notified the International Telecommunications Union (ITU) that it objects to such communications

Q. Under which of the following circumstances may an amateur radio station make one-way transmissions?

A. B. When transmitting code practice, information bulletins, or transmissions necessary to provide emergency communications

Q. When is it permissible to transmit messages encoded to hide their meaning?

A. Only when transmitting control commands to space stations or radio control craft

Q. Under what conditions is an amateur station authorized to transmit music using a phone emission?

A. When incidental to an authorized retransmission of manned spacecraft communications

Q. When may amateur radio operators use their stations to notify other amateurs of the availability of equipment for sale or trade?

A. When the equipment is normally used in an amateur station and such activity is not conducted on a regular basis

Q. What, if any, are the restrictions concerning transmission of language that may be considered indecent or obscene?

A. Any such language is prohibited

Q. What types of amateur stations can automatically retransmit the signals of other amateur stations?

A. Repeater, auxiliary, or space stations

Q. In which of the following circumstances may the control operator of an amateur station receive compensation for operating that station?

A. When the communication is incidental to classroom instruction at an educational institution

Q. Under which of the following circumstances are amateur stations authorized to transmit signals related to broadcasting, program production, or news gathering, assuming no other means is available?

A. Only where such communications directly relate to the immediate safety of human life or protection of property

[In my opinion, the question and answer above is misleading. Under 47 CFR §97.111(b)(6), "an amateur station may transmit … one-way communications … necessary to disseminate information bulletins." Along those lines, Amateur Radio Newsline states, "In the United States, ARNewsline can be re-transmitted by any licensed radio amateur as long as his/her station operation conforms to Sections 97.111 and 97.113 regarding one-way transmissions." Nevertheless, you need to select the listed answer.]

Q. What is the meaning of the term broadcasting in the FCC rules for the Amateur Radio Service?

A. Transmissions intended for reception by the general public

Q. When may an amateur station transmit without on-the-air identification?

A. When transmitting signals to control model craft

T1E - Control operator and control types: control operator required;

eligibility; designation of control operator; privileges and duties; control point; local, automatic and remote control; location of control operator

Q. When is an amateur station permitted to transmit without a control operator?

A. Never

Q. Who may be the control operator of a station communicating through an amateur satellite or space station?

A. Any amateur whose license privileges allow them to transmit on the satellite uplink frequency

Q. Who must designate the station control operator?

A. The station licensee

Q. What determines the transmitting privileges of an amateur station?

A. The class of operator license held by the control operator

Q. What is an amateur station control point?

A. The location at which the control operator function is performed

Q. When, under normal circumstances, may a Technician class licensee be the control operator of a station operating in an exclusive Amateur Extra class operator segment of the amateur bands?

A. At no time

Q. When the control operator is not the station licensee, who is responsible for the proper operation of the station?

A. The control operator and the station licensee are equally responsible

Q. Which of the following is an example of automatic control?

A. Repeater operation

Q. Which of the following is true of remote control operation?

A. A. The control operator must be at the control point
B. A control operator is required at all times
C. The control operator indirectly manipulates the controls
D. All of these choices are correct

Q. Which of the following is an example of remote control as defined in Part 97?

A. Operating the station over the internet

Q. Who does the FCC presume to be the control operator of an amateur station, unless documentation to the contrary is in the station records?

A. The station licensee

T1F - Station identification; repeaters; third-party communications; club stations; FCC inspection

Q. When must the station licensee make the station and its records available for FCC inspection?

A. At any time upon request by an FCC representative

Q. When using tactical identifiers such as "Race Headquarters" during a community service net operation, how often must your station transmit the station's FCC-assigned call sign?

A. At the end of each communication and every ten minutes during a communication

Q. When is an amateur station required to transmit its assigned call sign?

A. At least every 10 minutes during and at the end of a communication

Q. Which of the following is an acceptable language to use for station identification when operating in a phone sub-band?

A. The English language

Q. What method of call sign identification is required for a station transmitting phone signals?

A. Send the call sign using a CW or phone emission

Q. Which of the following formats of a self-assigned indicator is acceptable when identifying using a phone transmission?

A. KL7CC stroke W3
B. KL7CC slant W3
C. KL7CC slash W3
D. All of these choices are correct

Q. Which of the following restrictions apply when a non-licensed person is allowed to speak to a foreign station using a station under the control of a Technician class control operator?

A. The foreign station must be one with which the U.S. has a third-party agreement

Q. What is meant by the term Third Party Communications?

A. A message from a control operator to another amateur station control operator on behalf of another person

Q. What type of amateur station simultaneously retransmits the signal of another amateur station on a different channel or channels?

A. Repeater station

Q. Who is accountable should a repeater inadvertently retransmit communications that violate the FCC rules?

A. The control operator of the originating station

Q. Which of the following is a requirement for the issuance of a club station license grant?

A. The club must have at least four members

SUB-ELEMENT T2 - Operating Procedures - [3 Exam Questions - 3 Groups]

T2A - Station operation: choosing an operating frequency; calling another station; test transmissions; procedural signs; use of minimum power; choosing an operating frequency; band plans; calling frequencies; repeater offsets

Q. Which of the following is a common repeater frequency offset in the 2 meter band?

A. Plus or minus 600 kHz

Q. What is the national calling frequency for FM simplex operations in the 2 meter band?

A. 146.520 MHz

Q. What is a common repeater frequency offset in the 70 cm band?

A. Plus or minus 5 MHz

Q. What is an appropriate way to call another station on a repeater if you know the other station's call sign?

A. Say the station's call sign, then identify with your call sign

Q. How should you respond to a station calling CQ?

A. Transmit the other station's call sign followed by your call sign

Q. Which of the following is required when making on-the-air test transmissions?

A. Identify the transmitting station

Q. What is meant by "repeater offset?"

A. The difference between a repeater's transmit frequency and its receive frequency

Q. What is the meaning of the procedural signal "CQ"?

A. Calling any station

Q. What brief statement indicates that you are listening on a repeater and looking for a contact?

A. Your call sign

Q. What is a band plan, beyond the privileges established by the FCC?

A. A voluntary guideline for using different modes or activities within an amateur band

Q. What term describes an amateur station that is transmitting and receiving on the same frequency?

A. Simplex

Q. Which of the following is a guideline when choosing an operating frequency for calling CQ?

A. Listen first to be sure that no one else is using the frequency
B. Ask if the frequency is in use
C. Make sure you are in your assigned band

D. All of these choices are correct

T2B – VHF/UHF operating practices: SSB phone; FM repeater; simplex; splits and shifts; CTCSS; DTMF; tone squelch; carrier squelch; phonetics; operational problem resolution; Q signals

Q. What is the most common use of the "reverse split" function of a VHF/UHF transceiver?

A. Listen on a repeater's input frequency

Q. What term describes the use of a sub-audible tone transmitted along with normal voice audio to open the squelch of a receiver?

A. CTCSS

Q. If a station is not strong enough to keep a repeater's receiver squelch open, which of the following might allow you to receive the station's signal?

A. Listen on the repeater input frequency

Q. Which of the following could be the reason you are unable to access a repeater whose output you can hear?

A. Improper transceiver offset
B. The repeater may require a proper CTCSS tone from your transceiver
C. The repeater may require a proper DCS tone from your transceiver
D. All of these choices are correct

Q. What might be the problem if a repeater user says your transmissions are breaking up on voice peaks?

A. You are talking too loudly

Q. What type of tones are used to control repeaters linked by the Internet Relay Linking Project (IRLP) protocol?

A. DTMF

Q. How can you join a digital repeater's "talk group"?

A. Program your radio with the group's ID or code

Q. Which of the following applies when two stations transmitting on the same frequency interfere with each other?

A. Common courtesy should prevail, but no one has absolute right to an amateur frequency

Q. What is a "talk group" on a DMR digital repeater?

A. A way for groups of users to share a channel at different times without being heard by other users on the channel

Q. Which Q signal indicates that you are receiving interference from other stations?

A. QRM

Q. Which Q signal indicates that you are changing frequency?

A. QSY

Q. Why are simplex channels designated in the VHF/UHF band plans?

A. So that stations within mutual communications range can communicate without tying up a repeater

Q. Where may SSB phone be used in amateur bands above 50 MHz?

A. In at least some portion of all these bands

Q. Which of the following describes a linked repeater network?

A. A network of repeaters where signals received by one repeater are repeated by all the repeaters

T2C – Public service: emergency and non-emergency operations; applicability of FCC rules; RACES and ARES; net and traffic procedures; operating restrictions during emergencies

Q. When do the FCC rules NOT apply to the operation of an amateur station?

A. Never, FCC rules always apply

Q. What is meant by the term "NCS" used in net operation?

A. Net Control Station

Q. What should be done when using voice modes to ensure that voice messages containing unusual words are received correctly?

A. Spell the words using a standard phonetic alphabet

Q. What do RACES and ARES have in common?

A. Both organizations may provide communications during emergencies

Q. What does the term "traffic" refer to in net operation?

A. Formal messages exchanged by net stations

Q. Which of the following is an accepted practice to get the immediate attention of a net control station when reporting an emergency?

A. Begin your transmission by saying "Priority" or "Emergency" followed by your call sign

Q. Which of the following is an accepted practice for an amateur operator who has checked into a net?

A. Remain on frequency without transmitting until asked to do so by the net control station

Q. Which of the following is a characteristic of good traffic handling?

A. Passing messages exactly as received

Q. Are amateur station control operators ever permitted to operate outside the frequency privileges of their license class?

A. Yes, but only if necessary in situations involving the immediate safety of human life or protection of property

Q. What information is contained in the preamble of a formal traffic message?

A. The information needed to track the message

Q. What is meant by the term "check," in reference to a formal traffic message?

A. The number of words or word equivalents in the text portion of the message

Q. What is the Amateur Radio Emergency Service (ARES)?

A. Licensed amateurs who have voluntarily registered their qualifications and equipment for communications duty in the public service

SUB-ELEMENT T3 – Radio wave characteristics: properties of radio waves; propagation modes – [3 Exam Questions - 3 Groups]

T3A - Radio wave characteristics: how a radio signal travels; fading; multipath; polarization; wavelength vs absorption; antenna orientation

Q. What should you do if another operator reports that your station's

2 meter signals were strong just a moment ago, but now they are weak or distorted?

A. Try moving a few feet or changing the direction of your antenna if possible, as reflections may be causing multi-path distortion

Q. Why might the range of VHF and UHF signals be greater in the winter?

A. Less absorption by vegetation

Q. What antenna polarization is normally used for long-distance weak-signal CW and SSB contacts using the VHF and UHF bands?

A. Horizontal

Q. What can happen if the antennas at opposite ends of a VHF or UHF line of sight radio link are not using the same polarization?

A. Signals could be significantly weaker

Q. When using a directional antenna, how might your station be able to access a distant repeater if buildings or obstructions are blocking the direct line of sight path?

A. Try to find a path that reflects signals to the repeater

Q. What term is commonly used to describe the rapid fluttering sound sometimes heard from mobile stations that are moving while transmitting?

A. Picket fencing

Q. What type of wave carries radio signals between transmitting and receiving stations?

A. Electromagnetic

Q. Which of the following is a likely cause of irregular fading of signals received by ionospheric reflection?

A. Random combining of signals arriving via different paths

Q. Which of the following results from the fact that skip signals refracted from the ionosphere are elliptically polarized?

A. Either vertically or horizontally polarized antennas may be used for transmission or reception

Q. What may occur if data signals arrive via multiple paths?

A. Error rates are likely to increase

Q. Which part of the atmosphere enables the propagation of radio signals around the world?

A. The ionosphere

Q. How might fog and light rain affect radio range on the 10 meter and 6 meter bands?

A. Fog and light rain will have little effect on these bands

Q. What weather condition would decrease range at microwave frequencies?

A. Precipitation

T3B - Radio and electromagnetic wave properties: the electromagnetic spectrum; wavelength vs frequency; nature and velocity of electromagnetic waves; definition of UHF, VHF, HF bands; calculating wavelength

Q. What is the name for the distance a radio wave travels during one complete cycle?

A. Wavelength

Q. What property of a radio wave is used to describe its polarization?

A. The orientation of the electric field

Q. What are the two components of a radio wave?

A. Electric and magnetic fields

Q. How fast does a radio wave travel through free space?

A. At the speed of light

Q. How does the wavelength of a radio wave relate to its frequency?

A. The wavelength gets shorter as the frequency increases

Q. What is the formula for converting frequency to approximate wavelength in meters?

A. Wavelength in meters equals 300 divided by frequency in megahertz

Q. What property of radio waves is often used to identify the different frequency bands?

A. The approximate wavelength

Q. What are the frequency limits of the VHF spectrum?

A. 30 to 300 MHz

Q. What are the frequency limits of the UHF spectrum?

A. 300 to 3000 MHz

Q. What frequency range is referred to as HF?

A. 3 to 30 MHz

Q. What is the approximate velocity of a radio wave as it travels through free space?

A. 300,000,000 meters per second

T3C - Propagation modes: line of sight; sporadic E; meteor and auroral scatter and reflections; tropospheric ducting; F layer skip; radio horizon

Q. Why are direct (not via a repeater) UHF signals rarely heard from stations outside your local coverage area?

A. UHF signals are usually not reflected by the ionosphere

Q. Which of the following is an advantage of HF vs VHF and higher frequencies?

A. Long distance ionospheric propagation is far more common on HF

Q. What is a characteristic of VHF signals received via auroral reflection?

A. The signals exhibit rapid fluctuations of strength and often sound distorted

Q. Which of the following propagation types is most commonly associated with occasional strong over-the-horizon signals on the 10, 6, and 2 meter bands?

A. Sporadic E

Q. Which of the following effects might cause radio signals to be heard despite obstructions between the transmitting and receiving stations?

A. Knife-edge diffraction

Q. What mode is responsible for allowing over-the-horizon VHF and

UHF communications to ranges of approximately 300 miles on a regular basis?

A. Tropospheric ducting

Q. What band is best suited for communicating via meteor scatter?

A. 6 meter band

Q. What causes tropospheric ducting?

A. Temperature inversions in the atmosphere

Q. What is generally the best time for long-distance 10 meter band propagation via the F layer?

A. From dawn to shortly after sunset during periods of high sunspot activity

Q. Which of the following bands may provide long distance communications during the peak of the sunspot cycle?

A. 6 or 10 meter bands

Q. Why do VHF and UHF radio signals usually travel somewhat farther than the visual line of sight distance between two stations?

A. The Earth seems less curved to radio waves than to light

SUB-ELEMENT T4 - Amateur radio practices and station set-up – [2 Exam Questions - 2 Groups]

T4A – Station setup: connecting microphones; reducing unwanted emissions; power source; connecting a computer; RF grounding; connecting digital equipment; connecting an SWR meter

Q. What must be considered to determine the minimum current capacity needed for a transceiver power supply?

A. Efficiency of the transmitter at full power output
B. Receiver and control circuit power
C. Power supply regulation and heat dissipation
D. All of these choices are correct

Q. How might a computer be used as part of an amateur radio station?

A. For logging contacts and contact information
B. For sending and/or receiving CW
C. For generating and decoding digital signals
D. All of these choices are correct

Q. Why should wiring between the power source and radio be heavy-gauge wire and kept as short as possible?

A. To avoid voltage falling below that needed for proper operation

Q. Which computer sound card port is connected to a transceiver's headphone or speaker output for operating digital modes?

A. Microphone or line input

Q. What is the proper location for an external SWR meter?

A. In series with the feed line, between the transmitter and antenna

Q. Which of the following connections might be used between a voice transceiver and a computer for digital operation?

A. Receive audio, transmit audio, and push-to-talk (PTT)

Q. How is a computer's sound card used when conducting digital communications?

A. The sound card provides audio to the radio's microphone input and converts received audio to digital form

Q. Which of the following conductors provides the lowest impedance to RF signals?

A. Flat strap

Q. Which of the following could you use to cure distorted audio caused by RF current on the shield of a microphone cable?

A. Ferrite choke

Q. What is the source of a high-pitched whine that varies with engine speed in a mobile transceiver's receive audio?

A. The alternator

Q. Where should the negative return connection of a mobile transceiver's power cable be connected?

A. At the battery or engine block ground strap

T4B - Operating controls: tuning; use of filters; squelch function; AGC; transceiver operation; memory channels

Q. What may happen if a transmitter is operated with the microphone gain set too high?

A. The output signal might become distorted

Q. Which of the following can be used to enter the operating frequency on a modern transceiver?

A. The keypad or VFO knob

Q. What is the purpose of the squelch control on a transceiver?

A. To mute receiver output noise when no signal is being received

Q. What is a way to enable quick access to a favorite frequency on your transceiver?

A. Store the frequency in a memory channel

Q. Which of the following would reduce ignition interference to a receiver?

A. Turn on the noise blanker

Q. Which of the following controls could be used if the voice pitch of a single-sideband signal seems too high or low?

A. The receiver RIT or clarifier

Q. What does the term "RIT" mean?

A. Receiver Incremental Tuning

Q. What is the advantage of having multiple receive bandwidth choices on a multimode transceiver?

A. Permits noise or interference reduction by selecting a bandwidth matching the mode

Q. Which of the following is an appropriate receive filter bandwidth for minimizing noise and interference for SSB reception?

A. 2400 Hz

Q. Which of the following is an appropriate receive filter bandwidth for minimizing noise and interference for CW reception?

A. 500 Hz

Q. What is the function of automatic gain control, or AGC?

A. To keep received audio relatively constant

Q. Which of the following could be used to remove power line noise or ignition noise?

A. Noise blanker

Q. Which of the following is a use for the scanning function of an FM transceiver?

A. To scan through a range of frequencies to check for activity

SUB-ELEMENT T5 – Electrical principles: math for electronics; electronic principles; Ohm's Law – [4 Exam Questions - 4 Groups]

T5A - Electrical principles, units, and terms: current and voltage; conductors and insulators; alternating and direct current; series and parallel circuits

Q. Electrical current is measured in which of the following units?

A. Amperes

Q. Electrical power is measured in which of the following units?

A. Watts

Q. What is the name for the flow of electrons in an electric circuit?

A. Current

Q. What is the name for a current that flows only in one direction?

A. Direct current

Q. What is the electrical term for the electromotive force (EMF) that causes electron flow?

A. Voltage

Q. How much voltage does a mobile transceiver typically require?

A. About 12 volts

Q. Which of the following is a good electrical conductor?

A. Copper

Q. Which of the following is a good electrical insulator?

A. Glass

Q. What is the name for a current that reverses direction on a regular basis?

A. Alternating current

Q. Which term describes the rate at which electrical energy is used?

A. Power

Q. What is the unit of electromotive force?

A. The volt

Q. What describes the number of times per second that an alternating current makes a complete cycle?

A. Frequency

Q. In which type of circuit is current the same through all components?

A. Series

Q. In which type of circuit is current the same through all components?

A. Series

Q. In which type of circuit is voltage the same across all components?

A. Parallel

T5B - Math for electronics: conversion of electrical units; decibels; the metric system

Q. How many milliamperes is 1.5 amperes?

A. 1500 milliamperes

For the preceding question, note that the metric prefix "milli" = 1/1,000. Thus, move the decimal point 3 spaces to the right.

Q. What is another way to specify a radio signal frequency of 1,500,000 hertz?

A. 1500 kHz

For the preceding question, note that the metric prefix "kilo" (as in kilohertz) is a thousand times the unit of measurement. In this instance, move the decimal point 3 spaces to the left.

Q. How many volts are equal to one kilovolt?

A. One thousand volts

Q. How many volts are equal to one microvolt?

A. One one-millionth of a volt

Q. Which of the following is equal to 500 milliwatts?

A. 0.5 watts

Q. If an ammeter calibrated in amperes is used to measure a 3000-milliampere current, what reading would it show?

A. 3 amperes

Q. If a frequency display calibrated in megahertz shows a reading of 3.525 MHz, what would it show if it were calibrated in kilohertz?

A. 3525 kHz

Q. How many microfarads are equal to 1,000,000 picofarads?

A. 1 microfarad

Q. What is the approximate amount of change, measured in decibels (dB), of a power increase from 5 watts to 10 watts?

A. 3 dB

Q. What is the approximate amount of change, measured in decibels (dB), of a power decrease from 12 watts to 3 watts?

A. -6 dB

Q. What is the amount of change, measured in decibels (dB), of a power increase from 20 watts to 200 watts?

A. 10 dB

Q. Which of the following frequencies is equal to 28,400 kHz?

A. 28.400 MHz

Q. If a frequency display shows a reading of 2425 MHz, what frequency is that in GHz?

A. 2.425 GHz

T5C - Electronic principles: capacitance; inductance; current flow in circuits; alternating current; definition of RF; definition of polarity; DC power calculations; impedance

Q. What is the ability to store energy in an electric field called?

A. Capacitance

Q. What is the basic unit of capacitance?

A. The farad

Q. What is the ability to store energy in a magnetic field called?

A. Inductance

Q. What is the basic unit of inductance?

A. The Henry

Q. What is the unit of frequency?

A. Hertz

Q. What does the abbreviation "RF" refer to?

A. Radio frequency signals of all types

Q. A radio wave is made up of what type of energy?

A. Electromagnetic

Q. What is the formula used to calculate electrical power in a DC circuit?

A. Power (P) equals voltage (E) multiplied by current (I)

The preceding answer is easier to remember if you reorder the two properties to be multiplied, so that the equation reads $P = I \times E$ (i.e., "PIE"). The symbol used for voltage is "E" for "electromotive force." The symbol used for current is "I" which originates from the French phrase intensité du courant (current intensity). Using simple algebra allows you to reorder this formula to solve for any of the three variables: $I = P/E; E = P/I; P = I \times E$. Maybe thinking about French cuisine will help you remember the "PIE" formula. Bon appétit!

Q. How much power is being used in a circuit when the applied voltage is 13.8 volts DC and the current is 10 amperes?

A. 138 watts

Apply the PIE formula: P = I (13.8) x E (10) = 138.

Q. How much power is being used in a circuit when the applied voltage is 12 volts DC and the current is 2.5 amperes?

A. 30 watts.

Q. How many amperes are flowing in a circuit when the applied voltage is 12 volts DC and the load is 120 watts?

A. 10 amperes

Q. What is impedance?

A. A measure of the opposition to AC current flow in a circuit

Q. What is a unit of impedance?

A. Ohms

Q. What is the proper abbreviation for megahertz?

A. MHz

T5D – Ohm's Law: formulas and usage; components in series and parallel

Q. What formula is used to calculate current in a circuit?

A. Current (I) equals voltage (E) divided by resistance (R)

Q. What formula is used to calculate voltage in a circuit?

A. A. Voltage (E) equals current (I) multiplied by resistance (R)

Q. What formula is used to calculate resistance in a circuit?

A. Resistance (R) equals voltage (E) divided by current (I)

So, how can you remember this Ohm's Law formula: $E = I \times R$?
For starters, recall that "E" stands for electromotive force (voltage) and "I" is intensity (amperage).

To help illustrate the formula, we're going to travel on the **Electromotive** *Express (a locomotive), which rides on* **Iron Rails**. *There's your "E," "I," and "R." Iron and Rails get multiplied to equal Electromotive. To calculate either of the two lower variables, think of the line of the rails as a division sign: divide the "E" by either the "I" or the "R" to determine the remaining value. Thus, to calculate "I" (amps), divide "E" by "R" (E/R=I). To calculate "R" (resistance), divide "E" by "I" (E/I=R). It's easy—really. You will have the opportunity to practice with a number of the following questions.*

Q. What is the resistance of a circuit in which a current of 3 amperes flows through a resistor connected to 90 volts?

A. 30 ohms   *[E (90) / I (3) = R (30). Easy math!]*

Q. What is the resistance in a circuit for which the applied voltage is 12 volts and the current flow is 1.5 amperes?

A. 8 ohms

Q. What is the resistance of a circuit that draws 4 amperes from a 12-volt source?

A. 3 ohms

Q. What is the current in a circuit with an applied voltage of 120 volts and a resistance of 80 ohms?

A. 1.5 amperes

Q. What is the current through a 100-ohm resistor connected across 200 volts?

A. 2 amperes

Q. What is the current through a 24-ohm resistor connected across 240 volts?

A. 10 amperes

Q. What is the voltage across a 2-ohm resistor if a current of 0.5 amperes flows through it?

A. 1 volt

Q. What is the voltage across a 10-ohm resistor if a current of 1 ampere flows through it?

A. 10 volts

Q. What is the voltage across a 10-ohm resistor if a current of 2 amperes flows through it?

A. 20 volts

Q. What happens to current at the junction of two components in series?

A. It is unchanged

Q. What happens to current at the junction of two components in parallel?

A. It divides between them dependent on the value of the components

Q. What is the voltage across each of two components in series with a voltage source?

A. It is determined by the type and value of the components

Q. What is the voltage across each of two components in parallel with a voltage source?

A. The same voltage as the source

SUB-ELEMENT T6 – Electrical components; circuit diagrams; component functions – [4 Exam Questions - 4 Groups]

T6A - Electrical components: fixed and variable resistors; capacitors and inductors; fuses; switches; batteries

Q. What electrical component opposes the flow of current in a DC circuit?

A. Resistor

Q. What type of component is often used as an adjustable volume control?

A. Potentiometer

Q. What electrical parameter is controlled by a potentiometer?

A. Resistance

Q. What electrical component stores energy in an electric field?

A. Capacitor

Q. What type of electrical component consists of two or more conductive surfaces separated by an insulator?

A. Capacitor

Q. What type of electrical component stores energy in a magnetic field?

A. Inductor

Q. What electrical component usually is constructed as a coil of wire?

A. Inductor

Q. What electrical component is used to connect or disconnect electrical circuits?

A. Switch

Q. What electrical component is used to protect other circuit components from current overloads?

A. Fuse

Q. Which of the following battery types is rechargeable?

A. Nickel-metal hydride
B. Lithium-ion
C. Lead-acid gel-cell
D. All of these choices are correct

Q. Which of the following battery types is not rechargeable?

A. Carbon-zinc

T6B – Semiconductors: basic principles and applications of solid state devices; diodes and transistors

Q. What class of electronic components uses a voltage or current signal to control current flow?

A. Transistors

Q. What electronic component allows current to flow in only one direction?

A. Diode

Q. Which of these components can be used as an electronic switch or amplifier?

A. Transistor

Q. Which of the following components can consist of three layers of semiconductor material?

A. Transistor

Q. Which of the following electronic components can amplify signals?

A. Transistor

Q. How is the cathode lead of a semiconductor diode often marked on the package?

A. With a stripe

Q. What does the abbreviation LED stand for?

A. Light Emitting Diode

Q. What does the abbreviation FET stand for?

A. Field Effect Transistor

Q. What are the names of the two electrodes of a diode?

A. Anode and cathode

Q. Which of the following could be the primary gain-producing component in an RF power amplifier?

A. Transistor

Q. What is the term that describes a device's ability to amplify a signal?

A. Gain

T6C - Circuit diagrams; schematic symbols

Q. What is the name of an electrical wiring diagram that uses standard component symbols?

A. Schematic

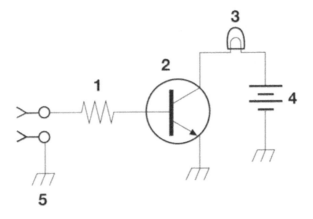

Figure T-1

Q. What is component 1 in figure T1?

A. Resistor

Q. What is component 2 in figure T1?

A. Transistor

Q. What is component 3 in figure T1?

A. Lamp

Q. What is component 4 in figure T1?

A. Battery

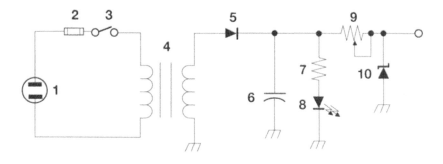

Figure T-2

Q. What is component 6 in figure T2?

A. Capacitor

Q. What is component 8 in figure T2?

A. Light Emitting Diode

Q. What is component 9 in figure T2?

A. Variable resistor

Q. What is component 4 in figure T2?

A. Transformer

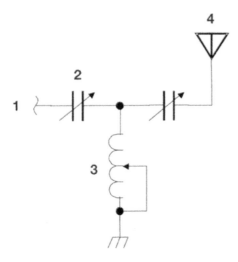

Figure T-3

Q. What is component 3 in figure T3?

A. Variable inductor

Q. What is component 4 in figure T3?

A. Antenna

Q. What do the symbols on an electrical schematic represent?

A. Electrical components

Q. Which of the following is accurately represented in electrical schematics?

A. The way components are interconnected

T6D - Component functions: rectification; switches; indicators; power supply components; resonant circuit; shielding; power transformers; integrated circuits

Q. Which of the following devices or circuits changes an alternating current into a varying direct current signal?

A. Rectifier

Q. What is a relay?

A. An electrically-controlled switch

Q. What type of switch is represented by component 3 in figure T2?

A. Single-pole single-throw

Q. Which of the following displays an electrical quantity as a numeric value?

A. Meter

Q. What type of circuit controls the amount of voltage from a power supply?

A. Regulator

Q. What component is commonly used to change 120V AC house current to a lower AC voltage for other uses?

A. Transformer

Q. Which of the following is commonly used as a visual indicator?

A. LED

Q. Which of the following is combined with an inductor to make a tuned circuit?

A. Capacitor

Q. What is the name of a device that combines several semiconductors and other components into one package?

A. Integrated circuit

Q. What is the function of component 2 in Figure T1?

A. Control the flow of current

Q. Which of the following is a resonant or tuned circuit?

A. An inductor and a capacitor connected in series or parallel to form a filter

Q. Which of the following is a common reason to use shielded wire?

A. To prevent coupling of unwanted signals to or from the wire

SUB-ELEMENT T7 – Station equipment: common transmitter and receiver problems; antenna measurements; troubleshooting; basic repair and testing – [4 Exam Questions - 4 Groups]

T7A – Station equipment: receivers; transmitters; transceivers; modulation; transverters; transmit and receive amplifiers

Q. Which term describes the ability of a receiver to detect the presence of a signal?

A. Sensitivity

Q. What is a transceiver?

A. A unit combining the functions of a transmitter and a receiver

Q. Which of the following is used to convert a radio signal from one frequency to another?

A. Mixer

Q. Which term describes the ability of a receiver to discriminate between multiple signals?

A. Selectivity

Q. What is the name of a circuit that generates a signal at a specific frequency?

A. Oscillator

Q. What device converts the RF input and output of a transceiver to another band?

A. Transverter

Q. What is meant by "PTT"?

A. The push-to-talk function that switches between receive and transmit

Q. Which of the following describes combining speech with an RF carrier signal?

A. Modulation

Q. What is the function of the SSB/CW-FM switch on a VHF power amplifier?

A. Set the amplifier for proper operation in the selected mode

Q. What device increases the low-power output from a handheld transceiver?

A. An RF power amplifier

Q. Where is an RF preamplifier installed?

A. Between the antenna and receiver

T7B – Common transmitter and receiver problems: symptoms of overload and overdrive; distortion; causes of interference; interference and consumer electronics; part 15 devices; over-modulation; RF feedback; off frequency signals

Q. What can you do if you are told your FM handheld or mobile transceiver is over-deviating?

A. Talk farther away from the microphone

Q. What would cause a broadcast AM or FM radio to receive an amateur radio transmission unintentionally?

A. The receiver is unable to reject strong signals outside the AM or FM band

Q. Which of the following can cause radio frequency interference?

A. Fundamental overload
B. Harmonics
C. Spurious emissions
D. All of these choices are correct

Q. Which of the following is a way to reduce or eliminate interference from an amateur transmitter to a nearby telephone?

A. Put an RF filter on the telephone

Q. How can overload of a non-amateur radio or TV receiver by an amateur signal be reduced or eliminated?

A. Block the amateur signal with a filter at the antenna input of the affected receiver

Q. Which of the following actions should you take if a neighbor tells you that your station's transmissions are interfering with their radio or TV reception?

A. Make sure that your station is functioning properly and that it does not cause interference to your own radio or television when it is tuned to the same channel

Q. Which of the following can reduce overload to a VHF transceiver from a nearby FM broadcast station?

A. Band-reject filter

Q. What should you do if something in a neighbor's home is causing harmful interference to your amateur station?

A. Work with your neighbor to identify the offending device
B. Politely inform your neighbor about the rules that prohibit the use of devices that cause interference
C. Check your station and make sure it meets the standards of good amateur practice
D. All of these choices are correct

Q. What is a Part 15 device?

A. An unlicensed device that may emit low-powered radio signals on frequencies used by a licensed service

Q. What might be a problem if you receive a report that your audio signal through the repeater is distorted or unintelligible?

A. Your transmitter is slightly off frequency
B. Your batteries are running low
C. You are in a bad location
D. All of these choices are correct

Q. What is a symptom of RF feedback in a transmitter or transceiver?

A. Reports of garbled, distorted, or unintelligible voice transmissions

Q. What should be the first step to resolve cable TV interference from your ham radio transmission?

A. Be sure all TV coaxial connectors are installed properly

T7C – Antenna measurements and troubleshooting: measuring SWR; dummy loads; coaxial cables; causes of feed line failures

Q. What is the primary purpose of a dummy load?

A. To prevent transmitting signals over the air when making tests

Q. Which of the following instruments can be used to determine if an antenna is resonant at the desired operating frequency?

A. An antenna analyzer

Q. What, in general terms, is standing wave ratio (SWR)?

A. A measure of how well a load is matched to a transmission line

Q. What reading on an SWR meter indicates a perfect impedance match between the antenna and the feed line?

A. 1 to 1

Q. Why do most solid-state amateur radio transmitters reduce output power as SWR increases?

A. To protect the output amplifier transistors

Q. What does an SWR reading of 4:1 indicate?

A. Impedance mismatch

Q. What happens to power lost in a feed line?

A. It is converted into heat

Q. What instrument other than an SWR meter could you use to determine if a feed line and antenna are properly matched?

A. Directional wattmeter

Q. Which of the following is the most common cause for failure of coaxial cables?

A. Moisture contamination

Q. Why should the outer jacket of coaxial cable be resistant to ultraviolet light?

A. Ultraviolet light can damage the jacket and allow water to enter the cable

Q. What is a disadvantage of air core coaxial cable when compared to foam or solid dielectric types?

A. It requires special techniques to prevent water absorption

Q. What does a dummy load consist of?

A. A non-inductive resistor and a heat sink

T7D – Basic repair and testing: soldering; using basic test instruments; connecting a voltmeter, ammeter, or ohmmeter

Q. Which instrument would you use to measure electric potential or electromotive force?

A. A voltmeter

Q. What is the correct way to connect a voltmeter to a circuit?

A. In parallel with the circuit

Q. How is a simple ammeter connected to a circuit?

A. In series with the circuit

Q. Which instrument is used to measure electric current?

A. An ammeter

Q. What instrument is used to measure resistance? A. An oscilloscope

A. An ohmmeter

Q. Which of the following might damage a multimeter?

A. Attempting to measure voltage when using the resistance setting

Q. Which of the following measurements are commonly made using a multimeter?

A. Voltage and resistance

Q. Which of the following types of solder is best for radio and electronic use?

A. Rosin-core solder

Q. What is the characteristic appearance of a cold solder joint?

A. A grainy or dull surface

Q. What is probably happening when an ohmmeter, connected across an unpowered circuit, initially indicates a low resistance and then shows increasing resistance with time?

A. The circuit contains a large capacitor

Q. Which of the following precautions should be taken when measuring circuit resistance with an ohmmeter?

A. Ensure that the circuit is not powered

Q. Which of the following precautions should be taken when measuring high voltages with a voltmeter?

A. Ensure that the voltmeter and leads are rated for use at the voltages to be measured

SUBELEMENT T8 – Modulation modes: amateur satellite operation; operating activities; non-voice and digital communications – [4 Exam Questions - 4 Groups]

T8A – Modulation modes: bandwidth of various signals; choice of emission type

Q. Which of the following is a form of amplitude modulation?

A. Single sideband

Q. What type of modulation is most commonly used for VHF packet radio transmissions?

A. FM

Q. Which type of voice mode is most often used for long-distance (weak signal) contacts on the VHF and UHF bands?

A. SSB

Q. Which type of modulation is most commonly used for VHF and UHF voice repeaters?

A. FM

Q. Which of the following types of emission has the narrowest bandwidth?

A. CW

Q. Which sideband is normally used for 10 meter HF, VHF, and UHF single-sideband communications?

A. Upper sideband

Q. What is an advantage of single sideband (SSB) over FM for voice transmissions?

A. SSB signals have narrower bandwidth

Q. What is the approximate bandwidth of a single sideband (SSB) voice signal?

A. 3 kHz

Q. What is the approximate bandwidth of a VHF repeater FM phone signal?

A. Between 10 and 15 kHz

Q. What is the typical bandwidth of analog fast-scan TV transmissions on the 70 centimeter band?

A. About 6 MHz

Q. What is the approximate maximum bandwidth required to transmit a CW signal?

A. 150 Hz

T8B - Amateur satellite operation; Doppler shift; basic orbits; operating protocols; transmitter power considerations; telemetry and telecommand; satellite tracking

Q. What telemetry information is typically transmitted by satellite beacons?

A. Health and status of the satellite

Q. What is the impact of using too much effective radiated power on a satellite uplink?

A. Blocking access by other users

Q. Which of the following are provided by satellite tracking programs?

A. Maps showing the real-time position of the satellite track over the earth
B. The time, azimuth, and elevation of the start, maximum altitude, and end of a pass
C. The apparent frequency of the satellite transmission, including effects of Doppler shift
D. All of these choices are correct

Q. What mode of transmission is commonly used by amateur radio satellites?

A. SSB
B. FM
C. CW/data
D. All of these choices are correct

Q. What is a satellite beacon?

A. A transmission from a satellite that contains status information

Q. Which of the following are inputs to a satellite tracking program?

A. The Keplerian elements

Q. With regard to satellite communications, what is Doppler shift?

A. An observed change in signal frequency caused by relative motion between the satellite and the earth station

Q. What is meant by the statement that a satellite is operating in mode U/V?

A. The satellite uplink is in the 70 centimeter band and the downlink is in the 2 meter band

The "U" stands for UHF (70 cm band) and the "V" for VHF (2 meter band).

Q. What causes spin fading of satellite signals?

A. Rotation of the satellite and its antennas

Q. What do the initials LEO tell you about an amateur satellite?

A. The satellite is in a Low Earth Orbit

Q. Who may receive telemetry from a space station?

A. Anyone who can receive the telemetry signal

Q. Which of the following is a good way to judge whether your uplink power is neither too low nor too high?

A. Your signal strength on the downlink should be about the same as the beacon

T8C – Operating activities: radio direction finding; radio control; contests; linking over the internet; grid locators

Q. Which of the following methods is used to locate sources of noise interference or jamming?

A. Radio direction finding

Q. Which of these items would be useful for a hidden transmitter hunt?

A. A directional antenna

Q. What operating activity involves contacting as many stations as possible during a specified period?

A. Contesting

Q. Which of the following is good procedure when contacting another station in a radio contest?

A. Send only the minimum information needed for proper identification and the contest exchange

Q. What is a grid locator?

A. A letter-number designator assigned to a geographic location

Q. How is access to some IRLP nodes accomplished?

A. By using DTMF signals

Q. What is meant by Voice Over Internet Protocol (VoIP) as used in amateur radio?

A. A method of delivering voice communications over the internet using digital techniques

Q. What is the Internet Radio Linking Project (IRLP)?

A. A technique to connect amateur radio systems, such as repeaters, via the internet using Voice Over Internet Protocol (VoIP)

Q. How might you obtain a list of active nodes that use VoIP?

A. By subscribing to an on line service
B. From on line repeater lists maintained by the local repeater frequency coordinator C. From a repeater directory
D. All of these choices are correct

Q. What must be done before you may use the EchoLink system to communicate using a repeater?

A. You must register your call sign and provide proof of license

Q. What name is given to an amateur radio station that is used to connect other amateur stations to the internet?

A. A gateway

T8D – Non-voice and digital communications: image signals; digital modes; CW; packet radio; PSK31; APRS; error detection and correction; NTSC; amateur radio networking; Digital Mobile/Migration Radio

Q. Which of the following is a digital communications mode?

A. Packet radio
B. IEEE 802.11
C. JT65
D. All of these choices are correct

Q. What does the term "APRS" mean?

A. Automatic Packet Reporting System

Q. Which of the following devices is used to provide data to the transmitter when sending automatic position reports from a mobile amateur radio station?

A. A Global Positioning System receiver

Q. What type of transmission is indicated by the term "NTSC?"

A. An analog fast scan color TV signal

Q. Which of the following is an application of APRS (Automatic Packet Reporting System)?

A. Providing real-time tactical digital communications in conjunction with a map showing the locations of stations

Q. What does the abbreviation "PSK" mean?

A. Phase Shift Keying

Q. Which of the following best describes DMR (Digital Mobile Radio)?

A. A technique for time-multiplexing two digital voice signals on a single 12.5 kHz repeater channel

Q. Which of the following may be included in packet transmissions?

A. A check sum that permits error detection
B. A header that contains the call sign of the station to which the information is being sent
C. Automatic repeat request in case of error
D. All of these choices are correct

Q. What code is used when sending CW in the amateur bands?

A. International Morse

Q. Which of the following operating activities is supported by digital mode software in the WSJT suite?

A. Moonbounce or Earth-Moon-Earth
B. Weak-signal propagation beacons
C. Meteor scatter
D. All of these choices are correct

Q. What is an ARQ transmission system?

A. A digital scheme whereby the receiving station detects errors and sends a request to the sending station to retransmit the information

Q. Which of the following best describes Broadband-Hamnet(TM), also referred to as a high-speed multi-media network?

A. An amateur-radio-based data network using commercial Wi-Fi gear with modified firmware

Q. What is FT8?

A. A digital mode capable of operating in low signal-to-noise conditions that transmits on 15-second intervals

Q. What is an electronic keyer?

A. A device that assists in manual sending of Morse code

SUB-ELEMENT T9 – Antennas and feed lines - [2 Exam Questions - 2 Groups]

T9A – Antennas: vertical and horizontal polarization; concept of gain; common portable and mobile antennas; relationships between resonant length and frequency; concept of dipole antennas

Q. What is a beam antenna?

A. An antenna that concentrates signals in one direction

Q. Which of the following describes a type of antenna loading?

A. Inserting an inductor in the radiating portion of the antenna to make it electrically longer

Q. Which of the following describes a simple dipole oriented parallel to the Earth's surface?

A. A horizontally polarized antenna

Q. What is a disadvantage of the "rubber duck" antenna supplied with most handheld radio transceivers when compared to a full-sized quarter-wave antenna?

A. It does not transmit or receive as effectively

Q. How would you change a dipole antenna to make it resonant on a higher frequency?

A. Shorten it

Q. What type of antennas are the quad, Yagi, and dish?

A. Directional antennas

Q. What is a disadvantage of using a handheld VHF transceiver, with its integral antenna, inside a vehicle?

A. Signals might not propagate well due to the shielding effect of the vehicle

Q. What is the approximate length, in inches, of a quarter-wavelength vertical antenna for 146 MHz?

A. 19

You do not need a calculator for the preceding question, as the answer can be estimated. 146 MHz is in the 2-meter band. One quarter length is a half meter, or about 1.5 feet (18 inches). None of the other multiple choice answers (112, 50, and 12) are particularly close.

Q. What is the approximate length, in inches, of a half-wavelength 6 meter dipole antenna?

A. 112

Again, you do not need a calculator for the preceding question, as the answer can be estimated. Half of a 6-meter wavelength is 3 meters. Since a meter is approximately 3 feet (39 inches), 3 meters is about 9 feet. Then convert to inches. If you don't know the multiplication table for factors of 12, round down to 10. 10 x 9 = 90 inches. Then take the 2 that you rounded off and multiply it by 9 to get 18. 90 + 18 = 108 inches. None of the other multiple choice answers (6, 50, and 236) are anywhere close to that.

Q. In which direction does a half-wave dipole antenna radiate the strongest signal?

A. Broadside to the antenna

Q. What is the gain of an antenna?

A. The increase in signal strength in a specified direction compared to a reference antenna

Q. What is an advantage of using a properly mounted 5/8 wavelength antenna for VHF or UHF mobile service?

A. It has a lower radiation angle and more gain than a 1/4 wavelength antenna

T9B – Feed lines: types, attenuation vs frequency, selecting; SWR concepts; Antenna tuners (couplers); RF Connectors: selecting, weather protection

Q. Why is it important to have low SWR when using coaxial cable feed line?

A. To reduce signal loss

Q. What is the impedance of most coaxial cables used in amateur radio installations?

A. 50 ohms

Q. Why is coaxial cable the most common feed line selected for amateur radio antenna systems?

A. It is easy to use and requires few special installation considerations

Q. What is the major function of an antenna tuner (antenna coupler)?

A. It matches the antenna system impedance to the transceiver's output impedance

Q. In general, what happens as the frequency of a signal passing through coaxial cable is increased?

A. The loss increases

Q. Which of the following connectors is most suitable for frequencies above 400 MHz?

A. A Type N connector

Q. Which of the following is true of PL-259 type coax connectors?

A. They are commonly used at HF frequencies

Q. Why should coax connectors exposed to the weather be sealed against water intrusion?

A. To prevent an increase in feed line loss

Q. What can cause erratic changes in SWR readings?

A. A loose connection in an antenna or a feed line

Q. What is the electrical difference between RG-58 and RG-8 coaxial cable?

A. RG-8 cable has less loss at a given frequency

Q. Which of the following types of feed line has the lowest loss at VHF and UHF?

A. Air-insulated hard line

SUBELEMENT T0 – Electrical safety: AC and DC power circuits; antenna installation; RF hazards – [3 Exam Questions - 3 Groups]

T0A – Power circuits and hazards: hazardous voltages; fuses and circuit breakers; grounding; lightning protection; battery safety; electrical code compliance

Q. Which of the following is a safety hazard of a 12-volt storage battery?

A. Shorting the terminals can cause burns, fire, or an explosion

Q. What health hazard is presented by electrical current flowing through the body?

A. It may cause injury by heating tissue
B. It may disrupt the electrical functions of cells
C. It may cause involuntary muscle contractions
D. All of these choices are correct

Q. In the United States, what is connected to the green wire in a three-wire electrical AC plug?

A. Equipment ground

Q. What is the purpose of a fuse in an electrical circuit?

A. To interrupt power in case of overload

Q. Why is it unwise to install a 20-ampere fuse in the place of a 5-ampere fuse?

A. Excessive current could cause a fire

Q. What is a good way to guard against electrical shock at your station?

A. Use three-wire cords and plugs for all AC powered equipment
B. Connect all AC powered station equipment to a common safety ground
C. Use a circuit protected by a ground-fault interrupter
D. All of these choices are correct

Q. Which of these precautions should be taken when installing devices for lightning protection in a coaxial cable feed line?

A. Mount all of the protectors on a metal plate that is in turn connected to an external ground rod

Q. What safety equipment should always be included in home-built equipment that is powered from 120V AC power circuits?

A. A fuse or circuit breaker in series with the AC hot conductor

Q. What should be done to all external ground rods or earth connections?

A. Bond them together with heavy wire or conductive strap

Q. What can happen if a lead-acid storage battery is charged or discharged too quickly?

A. The battery could overheat, give off flammable gas, or explode

Q. What kind of hazard might exist in a power supply when it is turned off and disconnected?

A. You might receive an electric shock from the charge stored in large capacitors

T0B – Antenna safety: tower safety and grounding; erecting an antenna support; safely installing an antenna

Q. When should members of a tower work team wear a hard hat and safety glasses?

A. At all times when any work is being done on the tower

Q. What is a good precaution to observe before climbing an antenna tower?

A. Put on a carefully inspected climbing harness(fall arrester) and safety glasses

Q. Under what circumstances is it safe to climb a tower without a helper or observer?

A. Never

Q. Which of the following is an important safety precaution to observe when putting up an antenna tower?

A. Look for and stay clear of any overhead electrical wires

Q. What is the purpose of a gin pole?

A. To lift tower sections or antennas

Q. What is the minimum safe distance from a power line to allow when installing an antenna?

A. Enough so that if the antenna falls unexpectedly, no part of it can come closer than 10 feet to the power wires

Q. Which of the following is an important safety rule to remember when using a crank-up tower?

A. This type of tower must not be climbed unless retracted or mechanical safety locking devices have been installed

Q. What is considered to be a proper grounding method for a tower?

A. Separate eight-foot long ground rods for each tower leg, bonded to the tower and each other

Q. Why should you avoid attaching an antenna to a utility pole?

A. The antenna could contact high-voltage power lines

Q. Which of the following is true when installing grounding conductors used for lightning protection?

A. Sharp bends must be avoided

Q. Which of the following establishes grounding requirements for an amateur radio tower or antenna?

A. Local electrical codes

Q. Which of the following is good practice when installing ground wires on a tower for lightning protection?

A. Ensure that connections are short and direct

Q. What is the purpose of a safety wire through a turnbuckle used to tension guy lines?

A. Prevent loosening of the guy line from vibration

T0C - RF hazards: radiation exposure; proximity to antennas; recognized safe power levels; exposure to others; radiation types; duty cycle

Q. What type of radiation are VHF and UHF radio signals?

A. Non-ionizing radiation

Q. Which of the following frequencies has the lowest value for Maximum Permissible Exposure limit?

A. 50 MHz

Q. What is the maximum power level that an amateur radio station may use at VHF frequencies before an RF exposure evaluation is required?

A. 50 watts PEP at the antenna

Q. What factors affect the RF exposure of people near an amateur station antenna?

A. Frequency and power level of the RF field
B. Distance from the antenna to a person
C. Radiation pattern of the antenna
D. All of these choices are correct

Q. Why do exposure limits vary with frequency?

A. The human body absorbs more RF energy at some frequencies than at others

Q. Which of the following is an acceptable method to determine that your station complies with FCC RF exposure regulations?

A. By calculation based on FCC OET Bulletin 65
B. By calculation based on computer modeling
C. By measurement of field strength using calibrated equipment
D. All of these choices are correct

Q. What could happen if a person accidentally touched your antenna while you were transmitting?

A. They might receive a painful RF burn

Q. Which of the following actions might amateur operators take to prevent exposure to RF radiation in excess of FCC- supplied limits?

A. Relocate antennas

Q. How can you make sure your station stays in compliance with RF safety regulations?

A. By re-evaluating the station whenever an item of equipment is changed

Q. Why is duty cycle one of the factors used to determine safe RF radiation exposure levels?

A. It affects the average exposure of people to radiation

Q. What is the definition of duty cycle during the averaging time for RF exposure?

A. The percentage of time that a transmitter is transmitting

Q. How does RF radiation differ from ionizing radiation (radioactivity)?

A. RF radiation does not have sufficient energy to cause genetic damage

Q. If the averaging time for exposure is 6 minutes, how much power density is permitted if the signal is present for 3 minutes and absent for 3 minutes rather than being present for the entire 6 minutes?

A. 2 times as much

End of Question Pool. Hurray! You made it through. You might want to take an online practice test while the information is fresh in your mind (unless your brain is tired and you feel like a zombie).

ABOUT THE AUTHOR

Tim Jacobson (call sign N9CD) has been a licensed amateur radio operator for only a short time, but he passed all three license exams on the same day, starting his ham radio adventure at the highest license class of "Amateur Extra" and dove into ham radio with great enthusiasm. He especially enjoys chasing DX and participating in contests. Through this book, he shares nuggets of information he gathered on his journey, with the goal of making it easier for others to get started in amateur radio.

Science and technology fascinate Jacobson. As a teen, he taught himself computer programming while also studying the natural world, and as a young adult he began building electronic circuits, including an AM radio, and started practicing CW using a Morse code tone generator he constructed.

During his college years, he worked as a blacksmith at a living history museum. Later, he earned a black belt in karate and served as a U.S. Air Force Auxiliary (Civil Air Patrol) mission pilot.

Jacobson works as an environmental attorney. Also, he served as executive producer of *Mysteries of the Driftless*, an Emmy Award-winning film of science exploration and adventure, and as a producer and cinematographer of *Decoding the Driftless*, another acclaimed documentary. He authored *The Kurchatov Penetration*, a techno-thriller that happens to mention the use of Morse code.

In addition to ham radio, Jacobson enjoys flying airplanes (land and seaplanes), biking and unicycling, landscape photography, travel, writing, playing guitar, rock climbing, scuba diving, and spending time with family. He lives along the scenic upper Mississippi River valley with his wife Lisa, not too far from his rambunctious, delightful grandkids.

Made in the USA
Middletown, DE
28 August 2021

47108408R00109